T0037977

Parched

Published in 2024 by Welbeck
An imprint of Welbeck Non-Fiction Limited
part of Welbeck Publishing Group
Offices in: London – 20 Mortimer Street, London W1T 3JW &
Sydney – Level 17, 207 Kent St, Sydney NSW 2000 Australia
www.welbeckpublishing.com

 A CIP catalogue for this book is available from the British Library.

ISBN 978-1-80279-725-1

Printed in Dubai

10 9 8 7 6 5 4 3 2 1

Parched

50 plants that survive and thrive in a dry garden

Philip Clayton

Contents

6 **Introduction**

8 **Silver wattle**
Acacia dealbata

12 **Bear's breeches**
Acanthus mollis

16 **Bell African lily**
Agapanthus campanulatus

20 **Century plant**
Agave americana

24 **Star of Persia**
Allium cristophii

28 **Barbados aloe**
Aloe vera

32 **Mexican lily**
Beschorneria yuccoides

36 **Bougainvillea**
Bougainvillea glabra

40 **Chinese trumpet vine**
Campsis grandiflora

44 **Californian lilac**
Ceanothus arboreus

46 **Fountain grass**
Cenchrus longisetus

50 **Judas tree**
Cercis siliquastrum

54 **Peruvian apple cactus**
Cereus repandus

58 **Dwarf fan palm**
Chamaerops humilis

62 **Sun rose**
Cistus × purpureus

66 **Pampas grass**
Cortaderia selloana

70 **Italian cypress**
Cupressus sempervirens

74 **Ivy-leaved cyclamen**
Cyclamen hederifolium

78 **Trailing ice plant**
Delosperma cooperi

82 **Tree poppy**
Dendromecon rigida

86 **African daisy**
Dimorphotheca jucunda (syn. *Osteospermum jucandum*)

90 **Mexican snowball**
Echeveria elegans

94 **California fuchsia**
Epilobium canum (syn. *Zauschneria californica*)

98 **Mexican fleabane**
Erigeron karvinskianus

102 **California poppy**
Eschscholzia californica

106 **Narrow-leaved black peppermint** *Eucalyptus nicholii*

110 **Mrs Robb's bonnet** *Euphorbia amygdaloides* subsp. *robbiae*

114 **Blue daisy** *Felicia amelloides*

118 **Giant fennel** *Ferula communis*

122 **Common fig** *Ficus carica*

126 **Mount Etna broom** *Genista aetnensis*

130 **Knotted cranesbill** *Geranium nodosum*

134 **Chapparal yucca** *Hesperoyucca whipplei*

138 **Algerian iris** *Iris unguicularis*

142 **English lavender** *Lavandula angustifolia*

146 **Lion's tail** *Leonotis leonurus*

150 **Catmint** *Nepeta racemosa*

154 **Bowden lily** *Nerine bowdenii*

158 **Oleander** *Nerium oleander*

162 **Olive** *Olea europaea*

166 **Prickly pear** *Opuntia ficus-indica*

170 **Rose-scented geranium** *Pelargonium capitatum*

174 **Canary Island date palm** *Phoenix canariensis*

178 **Stone pine** *Pinus pinea*

182 **Californian tree poppy** *Romneya coulteri*

186 **Beach rose** *Rosa rugosa*

190 **Common rosemary** *Salvia rosmarinus*

194 **Winter daffodil** *Sternbergia lutea*

198 **Purpletop vervain** *Verbena bonariensis*

202 **Lesser periwinkle** *Vinca minor*

204 **Index**

208 **Credits**

Introduction

Climate change is beginning to have an impact on many areas of our lives, with gardeners among the 'first responders' to its effects. Rainfall seems to be ever more unpredictable, winters generally milder and wetter, with summers seemingly drier and hotter. Drought or dry conditions are of course nothing new, with many gardeners experienced in fighting the conditions, battling with watering can and garden hose to succeed with the lush, leafy moisture-loving plants we often crave. Adding shelter to exposed sites and incorporating moisture retaining organic matter to soil can certainly help, but increasingly it's a never-ending battle that comes with financial and environmental costs.

Rather than fighting with garden conditions and a changing climate, gardeners should now consider adapting their plant choices, seeking out plants hailing from drier, warmer climates such as the Mediterranean, South America and Australia. This has been done before; in the 1960s Essex plantswoman Beth Chatto, helped by her ecologist husband Andrew, developed a beautiful experimental garden in one of the driest parts of the UK. Plants the couple suspected would thrive due to their wild origins were chosen, and through trial-and-error Beth helped change the way people look at dry gardens and what they chose to plant through her 'right plant, right place' ethos. Dry gardens are now a popular way of dealing with conditions once considered difficult. At the Royal Botanic Gardens, Kew, areas such as the Mediterranean Garden and the Duke's Garden beside Cambridge Cottage are filled with water-wise ideas and a wealth of drought-tolerant plants to inspire visitors.

Even in places where droughts have long been part of life, such as in Australia and South America, interest in often native plants that stand conditions better than introduced moisture-lovers is on the rise. In more temperate climates, part of the appeal of gardening with dry conditions rather than against them is that many great drought-resistant species make spectacular, easy garden plants. There is also a vast range to choose from, taken from the world's naturally dry environments, be they shimmering multi-hued *Cistus* from the Mediterranean, spectacular *Romneya* with huge poppy-like flowers from California, winter flowering *Acacia* from Australia, vibrant *Leonotis* from southern Africa or succulent *Agave* and of course cacti from the Americas. Some of these plants are hardy and stand cold temperate winters – especially when dry – others tolerate just a little cold, while the rest need to spend winter frost-free. Many thrive at the Royal Botanic Gardens, Kew helped by our drier, warmer summers, and these collections of plants are now helping scientists understand adaptations that make plants drought tolerant.

With a changing climate, rainfall is becoming ever more unpredictable and often a resource too precious to lavish on ill-chosen planting; even those long-popular garden plants that once flourished can, without attention, now often struggle. It is time for us to look carefully at what we grow and open our minds to the possibilities a drier climate can provide. The plants adapted long ago, and so now must we.

Silver wattle

Acacia dealbata

A native of South Eastern Australia and Tasmania, this fast-growing and spectacular evergreen tree is popular in mild, temperate districts for its impressive scented yellow winter flowers and delicate sea-green foliage. It tolerates light frosts and stands drought well once established.

The dazzling golden clouds of flowers produced by this Australian tree can make a remarkable and uplifting sight on dull winter days, transporting fortunate passers-by to the French Riviera on sight and attracting any pollinating bees brave enough to venture out. In cool climates it is a good tree for coastal or urban areas, where long, hard winters are rare; many a London street is enlivened by these arresting and seemingly unseasonal displays in January and February, flowers often covering the entire tree.

Wattles are members of the pea family (Fabaceae); this species – one of the hardiest, surviving temperatures around -5°C – is native to South Eastern Australia. The compact and hardier *A. dealbata* subsp. *subalpina* grows at altitude in the Snowy Mountains of southern New South Wales. Silver wattle is a pioneer species – one of the first plants to recolonize areas of bush after wildfires as a result of its quick-growing habit. It is not long-lived, and plants seldom exceed 30–40 years. It was introduced to the UK in 1820 and was grown initially under glass; in southern Europe – especially the south of France – it became popular and still is grown as a cut flower. Although beautiful, it has become a menace in some areas, escaping into the wild where it can form large expanses of woodland; in parts of the Mediterranean and countries such as South Africa and New Zealand, it is classed as an invasive species.

Opposite Silver wattle (*Acacia dealbata*) from *The Garden*, 1892.

Silver wattle (also known as blue wattle or mimosa) makes a rewarding garden plant for cooler areas, as it develops remarkably quickly when happy, making a single or multi-stemmed tree, with smooth grey bark. Initially plants are upright and slender but with age become spreading, forming a tree with a rounded crown to around 25m, often wider than tall. The bipinnate evergreen foliage is appealing – young shoots protected by silvery hairs, leaves 12–15cm long and composed of tiny individual leaflets, fern-like and silvery-green in hue. Flowers begin to develop late in the previous summer at the tips of new shoots, most profusely during a hot, sunny season. The tiny individual blooms composed mostly of stamens are held in ball-like heads, which are in turn carried in panicles up to 12cm long. They may be followed by flat silvery seed pods.

Opposite Silver wattle (*Acacia dealbata*) from *Floral Cabinet and Magazine of Exotic Botany* by George Beauchamp Knowles and Frederic Westcott, 1840.

Plant silver wattle in full sun, somewhere with space; by a south-facing wall is good but at least 2m away. Soil needs to be well drained. Poor sandy or stony soil, acid or alkaline, will suit it, but this tree does not like shallow chalk. Choose a place sheltered from the wind: its evergreen crown means the tree is easily damaged by winter gales; young trees should be supported by a tree stake until established. The selection 'Gaulois Astier' is sometimes available and said to be a more compact plant, but if your wattle gets too big, it is possible to prune after flowering in spring – any later and you'll lose next year's flowers. Ferocious winters can kill plants or knock them back to the trunk – shoots may appear or sometimes suckers sprout from the roots to regenerate. It's always worth cutting some branches just as the tree starts to flower and bringing them inside; they last well in water and scent the house beautifully. The roots of this tree can be fairly shallow, so planting beneath may be challenging – although helpfully the tree's crown generally casts only light shade; choose drought-tolerant plants such *Euphorbia characias*, *Vinca difformis*, *Geranium endressii* and *Cyclamen hederifolium*, as well as spring bulbs. Nearby, out in the sun, plants such as lavenders, *Cistus* and *Melaleuca* all associate well.

OTHER *ACACIA* TO TRY

- ***Acacia baileyana*** (Cootamundra wattle) needs a sheltered place, where it will make a small rounded evergreen tree to 6m, flowering in early spring, yellow blooms held in short racemes. The selection 'Purpurea' has remarkable purple-blue foliage. Keep above -5°C.
- ***Acacia pravissima*** (Oven's wattle) is a great choice for a sunny wall. It is quite different in appearance from ***A. dealbata*** with arching branches and small triangular blue-green leaf-like phyllodes (winged leaf stalks). Clusters of showy yellow flowers appear in spring.

Bear's breeches

Acanthus mollis

Forming ground-covering mounds of lush, glossy rich green, often deeply lobed foliage, this hardy perennial has long been admired for its handsome appearance, especially in summer when impressive spires of white and purple-pink flowers arise, often attracting bumblebees.

Few plants deserve the description 'architectural' better than this bold, long-cultivated perennial; it was probably enjoyed in ancient Greek and Roman gardens and it is the stylized foliage of this noble plant which famously adorns the capitals of Corinthian columns, employed widely in classical architecture and furniture for centuries. It is certainly a plant with impact; well grown, it can be impressive for much of the year, initially due to that fabulous foliage, then later in summer with its terrific spikes of white and purple flowers.

Native to countries bordering the Mediterranean, it is well adapted to dry conditions, with fleshy, tuberous roots that spread strongly below ground, forming rafts of growth and offering, in gardens, first-rate ground cover. In the wild it grows in woods and rocky areas, usually in some shade. In some warmer countries such as South Africa and New Zealand, its spreading habit is problematic and invasive. In gardens it is usually pretty easily controlled.

Plants in cultivation vary. The most impressive are often described horticulturally as *A. mollis* (Latifolius Group): these are examples with the largest and broadest leaves, to around 1m or more long, with shallow lobes. In spring this fresh new growth uncoils from the ground – bright green, smooth and wonderfully glossy, with a shine like patent leather. Frost may occasionally cause damage, but such is the plant's vigour that burned leaves are soon masked by lavish rosettes of soft, arching leaves forming mounds of growth often to 3m across. Then in summer, flowering stems arise. These are slightly prickly to the touch and most numerous following a previously warm season. They reach around 1m or more tall and create an impressive display. Individual flowers are white or pink-tinged, each with contrasting, usually purple-pink hood-like bracts and well-spaced up the spike, opening from the bottom upwards. The flowers are usually well attended by bumblebees and, after flowering, rounded capsules containing the seeds form in the spikes if they are allowed to remain in place. The purple bracts are responsible for this plant's common name; they are said to resemble a bear's claw.

Opposite Bear's breeches (*Acanthus mollis*) from *The New Botanic Garden* by Sydenham Teast Edwards, 1812.

Pl. VII.
h
b
d
f
c
e
a
g
J. Jefferys sculp.
ACANTHUS mrioribus & bre.rioribus, aculeis munitis: Tourn. Inst. 176.
Published according to the Act by P. Miller April 29. 1755.

Opposite Bear's breeches (*Acanthus mollis*) from *Figures of the Most Beautiful, Useful, and Uncommon Plants Described in the Gardeners Dictionary* by Philip Miller and Georg Dionysius Ehret, 1755.

By late summer, especially in a hot very dry year, plants can start to look messy, the leaves limp and damaged by slugs and snails – and, occasionally, powdery mildew can strike plants. It is best in these circumstances to crop off the worst foliage – fresh leaves will arise in the cooler, moist conditions of autumn and may stay looking good until the first hard ground frosts; in mild city gardens, plants often prove virtually evergreen.

When planting *Acanthus mollis*, choose ideally a warm, sunny place, although it stands light shade and will even grow in north-facing borders. It is not too fussy about soil as long as well drained, and when established will be a long-lasting garden plant which will need little attention for years, although a spring mulch of garden compost will help keep the plant in good growth. This perennial usually looks best with plenty of space – the corner of a large herbaceous border, in a gravel garden or at the sunny edge of a woodland border. It looks super beside stone or brickwork or around the base of statuary for a classic look. Teaming with other plants is tricky because the running roots and lavish foliage usually outcompete neighbours, but it looks great as a contrast to spiky *Yucca gloriosa* or seen with other bold foliage such as that of palm *Chamaerops humilis*. It also grows freely from bits of root left in the ground after division or removal of unwanted runners, so bear this in mind if clearing it from a border.

OTHER *ACANTHUS* TO TRY

- ***Acanthus mollis* 'Hollard's Gold'** is a superb selection which has startling glossy buttercup-yellow foliage in spring and again in autumn, ageing to Chartreuse-green and darker as the season progresses. It is rather less vigorous and looks remarkable beside purple tulips.
- ***A. mollis*** (Latifolius Group) **'Rue Ledan'** has rather different flowers – the purple-pink bracts in this case replaced with green, giving the plant a rather silvery look when in flower.
- ***A. spinosus*** is in some ways even more appealing with deeply cut, surprisingly spiny, dark-green white-veined foliage to around 60cm long. The flowers are similar to ***A. mollis*** and appear after a hot summer.

Bell African lily

Agapanthus campanulatus

Among the most beautiful of perennials, *Agapanthus* thrive best in hot, sunny summers when they can provide a sensational display, producing rounded heads of blooms, in blue or white, held high above foliage. They make great plants for containers too.

Opposite Bell African lily (*Agapanthus campanulatus*) illustrated by Malcolm English, 2024.

With a succession of impressive rounded heads, composed of masses of usually blue or white bell-shaped flowers, sun-loving *Agapanthus* are terrific in summer. The recent rise of these perennials as popular garden plants in temperate regions is due to two factors: hotter, drier summers, which mean quite tender selections can now be relied upon to perform well, and breeding work, which has delivered superb plants with long flowering seasons, bigger flower heads and a broader range of colours to choose from, not only traditional white or blue, but also bicolours, mauves and purples.

Despite these excitements, *Agapanthus campanulatus* remains among the most dependable for temperate gardens, introduced to the UK around 1870. It is deciduous (some more evergreen kinds are tender) and regarded as one of the hardiest, standing winter temperatures of -10°C. The species is native to the Drakensberg Mountains in eastern South Africa, where it grows in grassland and rocky valleys. It is a clump-forming plant, with dark green, strap-shaped leaves to 40cm long, emerging from a fleshy, slowly spreading rhizome which helps it resist dry periods, notably when dormant in winter. Individual flowers are bell-shaped and up to 3cm long, arranged in flat-topped heads composed of dozens of blooms, lasting weeks in beauty, held atop stout stems to around 1m.

In the garden it is easy-going in the right conditions – sun is the key requirement in temperate areas, so choose as bright a place as possible. Plants revel in the reflected heat of a sunny wall if planted nearby, and ensure any neighbouring plants do not cast shade on your plant. *Agapanthus* generally are drought-tolerant, and more likely to make it through exceptionally long, dry summers than many other popular perennials. Flowering performance in the driest seasons will, however, be reduced unless some water can be provided, especially for plants in containers. Although generally needing dry conditions, when dormant in winter, *A. campanulatus* will tolerate some wet weather as long as the soil is free draining.

Agapanthus make great choices for the front of borders – foliage is low while the flower heads are carried on slender

Opposite Blue lily (*Agapanthus praecox*) from *Edwards's Botanical Register*, 1843.

stems, allowing views through to other plants. They can be ideal for gravel gardens where the plant's elegant overall form is easily admired; smaller selections thrive on rock gardens. Plants flourish on sunny slopes, banks and wall tops where they enjoy excellent drainage. Bumblebees are frequent visitors. You can cut the flowers – they last several weeks in a vase.

In especially cold or waterlogged locations, *Agapanthus* are best kept in containers and can look superb, flowering freely. This is also a good way to keep more tender evergreen ones; you can tuck them into a greenhouse or cold frame over winter. Hardier deciduous kinds are fine in a shed or even by a sheltered house wall. You'll need to feed potted plants regularly.

Planting partners need consideration – it is important not to let *Agapanthus* become crowded out by neighbours. Slender ornamental grasses such as *Stipa tenuissima* and bulbous plants, particularly *Crocosmia*, seem to do well beside them, while *Erigeron karvinskianus* (Mexican fleabane) makes super underplanting. Planted nearby, *Salvia*, *Nepeta*, *Lavandula* and late daisies such as *Rudbeckia* can all make great choices.

After flowering, allow *A. campanulatus* foliage to yellow – you can let the faded flower heads stand through winter because they can look attractive dusted with frost, and allowing seeds to fall can result in occasional seedlings popping up. Apply a mulch of garden compost over the plant's crown to protect from winter cold – this will help feed plants too.

OTHER *AGAPANTHUS* TO TRY

- ***A.* 'Alan Street'** produces intense dark violet-purple flowers held atop 1m tall stems. Deciduous and hardy to -10°C.
- ***A. inapertus*** proves distinctive, its nodding, tubular blue flowers hanging in small heads from tall 1.4m stems. Deciduous and hardy to -10°C.
- ***A.* Twister ('Ambic001')** has large heads of pure white flowers with a blue throat creating a bicoloured effect. Semi-evergreen, hardy to -10°C.
- ***A.* Poppin' Purple** (**'MP003'**) has large heads of violet-purple flowers. It is evergreen and hardy to -5°C, so best for a pot with winter protection.
- ***A.* 'Windsor Grey'** has flowers of pale grey-blue held on 1m stems. Deciduous and hardy to -10°C.
- ***A. praecox*** is a variable evergreen species to 1m, and a parent of many cultivars. Hardy to -5c.

Century plant

Agave americana

Forming impressive rosettes of spine-tipped succulent leaves, this drought-proof plant makes a great feature in a container, dry garden or bedded out for summer; some hardier species can be kept outdoors for year-round effect even in frost-prone gardens.

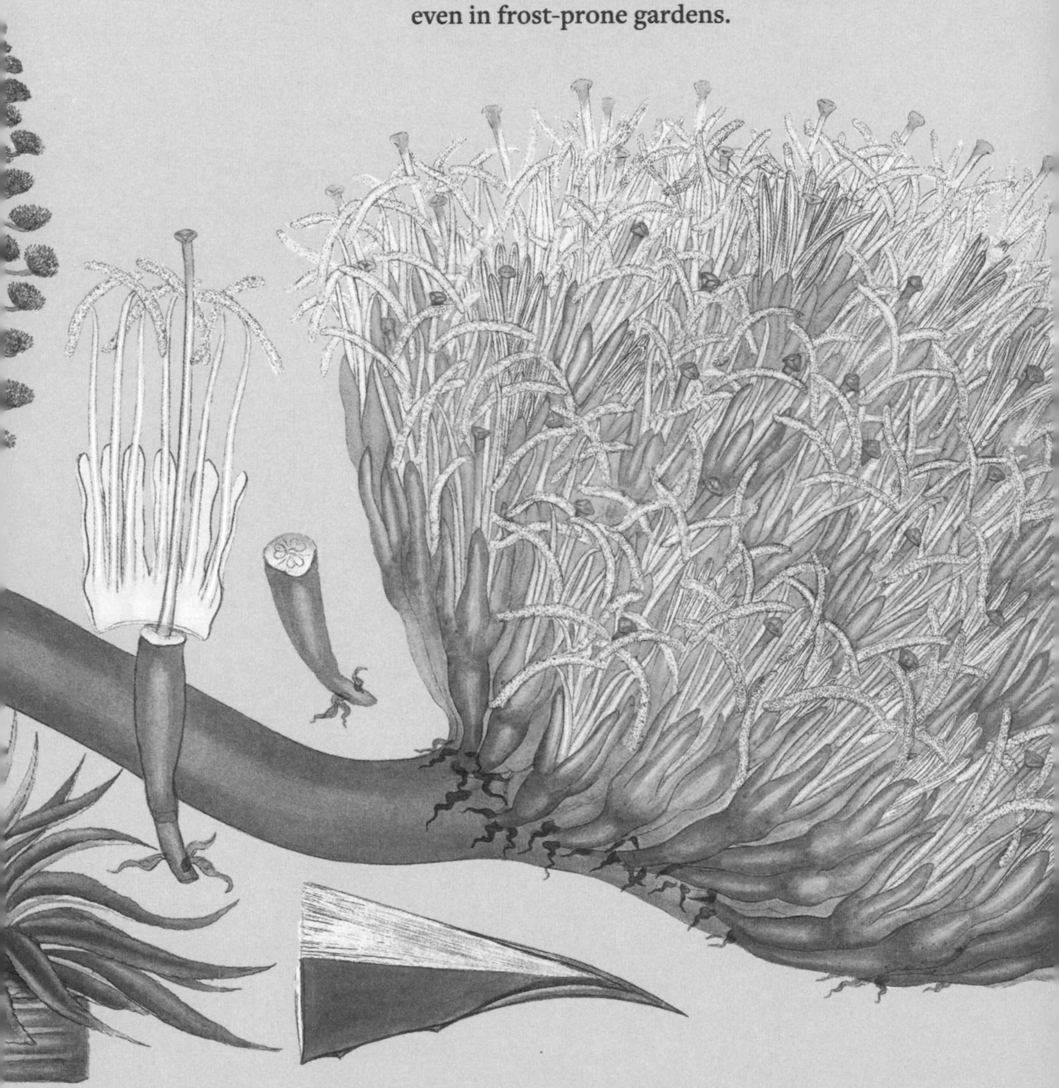

It is hard to think of a plant more obviously adapted for surviving drought than *Agave*; a genus of around 270 species of usually boldly succulent, rosette-forming, monocarpic (one-time flowering) slow-growing perennials, with fleshy, often spine-tipped and edged leaves. These arresting plants are native to arid parts of the New World – North America, Mexico and the Caribbean – but some, particularly impressive *A. americana* are widely cultivated in dry regions worldwide, and can be found naturalized in, for example, many Mediterranean countries. Some species are of economic importance: *A. azul* (blue agave) is used to make tequila and is the main source of agave nectar; the foliage of *A. sisalana* contains fibres used to make sisal. Many, though, are grown for their ornamental qualities, both in gardens and under glass. Most are tender in temperate climes and often bedded out in summer, only to return to greenhouses for a frost-free winter; a few species have proved hardy in recent years, however, especially in mild coastal or city gardens.

Opposite Century plant (*Agave americana*) from *The Botanist's Repository for New and Rare Plants* by Henry Charles Andrews, 1816.

Introduced to Europe from tropical America in the sixteenth century, *A. americana* is probably the best-known in cultivation. Its common name, century plant, derives from the mistaken belief that plants take 100 years to produce flowers. Certainly, it can prove slow-growing, especially in poor conditions, but generally plants flower after 10–12 years, and usually before they reach the ripe old age of 30; other species may take 60 years to produce a flowering spike. The flowering of *A. americana* is distinctly spectacular, the plant producing a branched, tree-like stem to 8m tall from the centre of its rosette. Side branches bear thousands of tubular yellow flowers in brush-like heads. After flowering, the main rosette of the plant dies, but not before dozens of offsets have formed around the base. Some species even produce young plantlets on the flowering spike – a belt and braces way of ensuring the plant endures.

While this species can flower in the mildest temperate gardens, it is usually kept for its architectural form; it certainly shrugs off drought with little consequence, surviving if need be with next to no water all summer. The thick, blue-green leaves make handsome rosettes; there are several frequently grown

Opposite Century plant (*Agave americana*) from *Revue horticole*, 1862.

variegated selections that are even more ornamental, such as 'Marginata' with gold-edged leaves or 'Mediopicta Alba' with a broad white stripe down the centre of each leaf. While the spines on the serrated leaf edges are sharp enough, the terminal needle at the end of each leaf is vicious; it is a wise precaution to cut the tip off with secateurs.

Usually, plants are kept in containers and displayed alongside other succulents, such as *Aloe* or various cacti. Alternatively, they can be bedded out in the hottest, sunniest places in the garden during summer, then moved into frost-free accommodation for winter – young plants will fit on a windowsill, but larger potted plants need dragging into a glasshouse or conservatory before wet, cold conditions strike. In fact, this plant can briefly tolerate temperatures of -5°C if grown in sharply drained soil – perhaps in a raised bed or rock garden filled with almost pure gravel, or perhaps on a hot, dry bank. Plants must stay dry in winter; moisture will rot the plant in low temperatures, so overhead protection is needed. The best plan is to remove and grow on offsets as they appear and experiment with planting these outdoors, while keeping the parent plant in a pot; other species, however, are hardier and may be more suited to year-round outside cultivation.

OTHER *AGAVE* TO TRY

- ***Agave attenuata*** (foxtail agave) is a most handsome species with smooth blue leaves, suited to growing in a pot. As it is spineless, handling is easy. The cylindrical, arching flower spike is spectacular too. Keep frost-free; stands light shade.
- ***A. montana*** (mountain agave) is one of the hardiest and a handsome plant. It comes from high-altitude pine forest in Mexico and forms a star-shaped rosette of fiercely toothed grey-green leaves, each with a rapier-like red-tinted terminal spine. It stands some winter wet and temperatures below -5°C when mature. It does not bear offsets.
- ***A. parryi*** (Parry's agave) is another to try outdoors – smaller than ***A. americana***, it has handsome silver-blue leaves and black spines forming a rounded rosette. It freely produces offsets and these are worth experimenting with around the garden, plants standing -5°C and tolerating some winter wet.

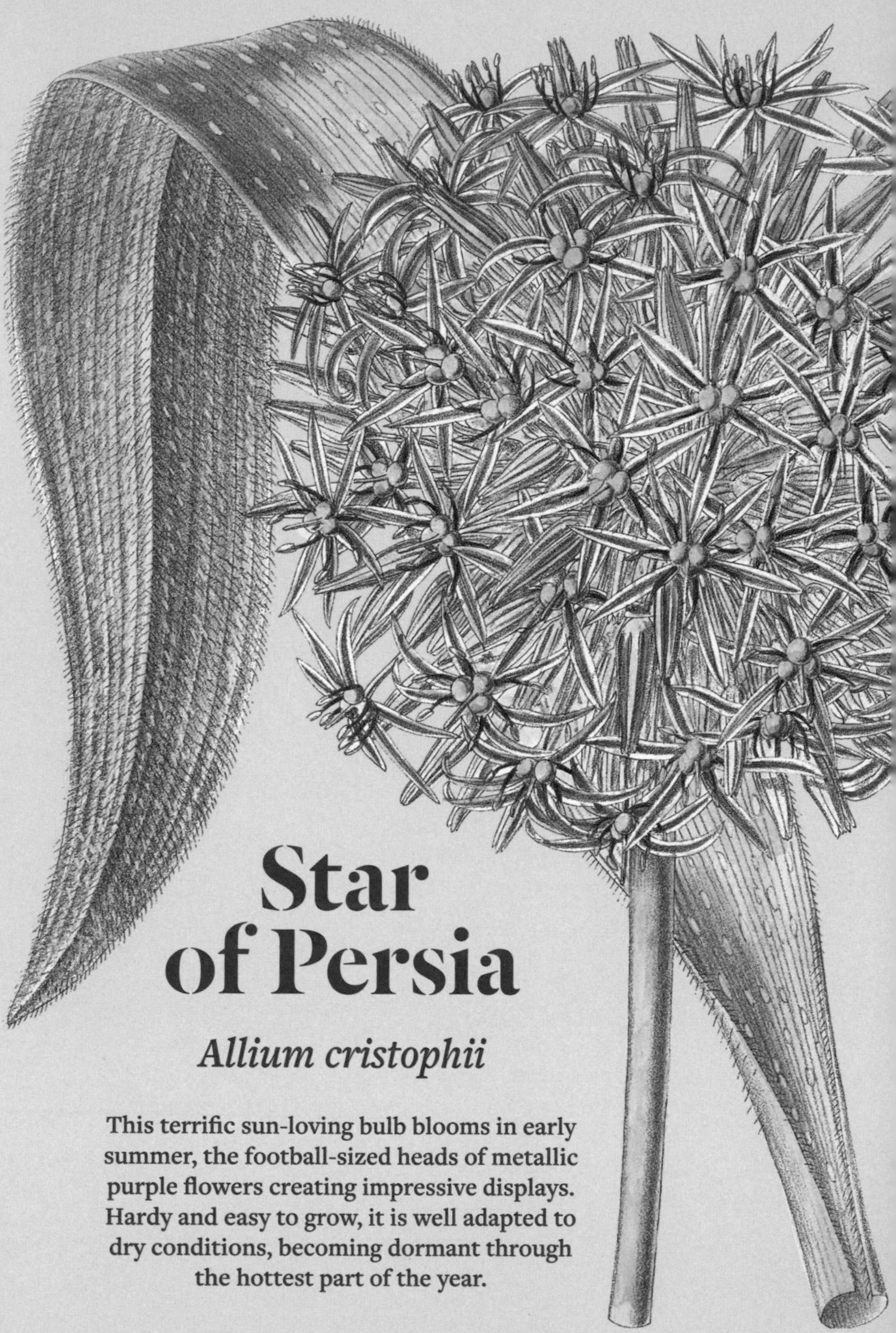

Star of Persia

Allium cristophii

This terrific sun-loving bulb blooms in early summer, the football-sized heads of metallic purple flowers creating impressive displays. Hardy and easy to grow, it is well adapted to dry conditions, becoming dormant through the hottest part of the year.

Alliums or ornamental onions were not widely grown in gardens 40 years ago, but from the 1990s onwards their popularity suddenly exploded, and now these often-bulbous perennials have become a staple plant of many temperate gardens. They bring colour and distinctive form to late spring and early summer plantings, filling a gap left by now-faded tulips and daffodils. Part of their great appeal is the way the flowers are arranged – in the case of some of the most popular kinds, in globular heads perched atop stems that allow them to combine well with other plants. All have that distinctive onion odour if crushed. Many are hardy and reliably drought-tolerant, and lovers of hot, sunny sites, enjoying well-drained soil. A few will spread freely, in some cases all too freely; others, such as splendid *Allium cristophii*, are well-behaved.

Native to the mountainous border between Iran and Turkmenistan, this fine species was introduced to Western gardens around the turn of the twentieth century. It has among the largest flower heads of all; wondrous head-sized globes of metallic silver-violet starry flowers, each individual bloom 3cm across. The heads are held on comparatively short stems – perhaps 45cm tall at most, often much less – and flower stems begin to arise just as the strappy foliage is beginning to die off, a characteristic of many alliums. The simplest thing to do is to remove the unsightly leaves – it does not harm the bulb once they are dying down. Flowers are long-lasting in beauty and turn to parchment brown as they become seedheads, lasting well through summer and into autumn, increasing the ornamental appeal. Allowing them to remain in situ also has the happy consequence of allowing plants to self-seed, which they do freely when happy.

Opposite Star of Persia (*Allium cristophii*) from *Curtis's Botanical Magazine*, 1904.

Plants are usually sold as dormant bulbs in autumn when they should be planted; they begin to emerge in early spring. This allium is highly tolerant of summer drought because it becomes dormant just as the heat of the season begins to reach its peak. Choose well-drained, ideally rather sandy soil to grow them in, and a sunny, fairly open site – bulbs are planted 10–15cm deep but do allow plenty of space in between to allow for the flower heads to expand.

Opposite Schubert's allium (*Allium schubertii*) from *Curtis's Botanical Magazine*, 1898.

The real trick is knowing what to plant them with for best effect – the short stems mean that the heads do not float obligingly mid-border atop other herbaceous plants, unlike some other alliums, so these are more suited for front-line displays or in lower, more naturalistic, less structured plantings. Planting *Allium* bulbs between rhizomes of tall bearded iris works well; they flower with the iris yet at a lower level, they cast little shade and after flowering the allium seedhead still provides interest; both plants enjoy similar conditions, bearded iris can be tricky to place amid other plants.

This allium also intermingles well with some smaller drought-tolerant ornamental grasses – it is super between clumps of *Festuca glauca* or *F. amethystina*, with *Nassella tenuissima* or billowing bronze-tinted *Anemanthele lessoniana*, partly as the grass foliage helps mask yellowing allium leaves. You could try them in front of taller roses – shrub roses or even hybrid teas and floribundas, all of which can have ungainly legs that are best covered, or in a gravel garden. Let them explode from a carpet of low-growing thyme, succulent *Petrosedum rupestre* or pink-flowered *Armeria maritima* (thrift) perhaps.

Once you have enough allium bulbs in your borders, lift and plant some in a cut flower border; the heads last two to three years indoors. Cut while flowers are fresh and hang up to dry. They look extraordinary in a vase, providing a memory of summer.

OTHER *ALLIUM* TO TRY

- ***Allium hollandicum* 'Purple Sensation'** (Dutch garlic) is probably the best-known ornamental onion with rich mauve flowers in dense round heads atop 70cm stems. It does well in sun or semi-shade, flowering in May. There are many fine hybrids involving this species with flowers in white and varying shades of mauve.
- ***A. schubertii*** (Schubert's allium) is an extraordinary sight in flower, the large heads of pale mauve flowers on 40cm stems looking something like a starburst, individual blooms making up the head on stalks of different lengths. Give it sharp drainage and full sun.
- ***A.* 'Summer Beauty'** blooms usefully later than most – usually in July – and is a compact clump-forming plant to 40cm with slender foliage that stays in good order for flowering. It has round, soft pink heads of flowers.

Barbados aloe

Aloe vera

Widely cultivated for commercial purposes in tropical regions, this handsome and useful succulent plant makes an easily grown houseplant in frost-prone areas that can even be stood outside for summer.

It seems that almost every commodity from health drinks and face creams to soap and even toilet paper now contains extracts from this invaluable plant. Its scientific name is surely one of the most recognized of all and has been widely grown in tropical and arid regions around the globe for around 400 years, for both ornamental and commercial purposes. In recent years the plant itself has even become a popular indoor choice in many temperate regions, grown for its ease of cultivation, handsome appearance and the soothing properties of its sap (see note below).

Native to desert and dry shrubland of northern Oman, it has become naturalized and even invasive in many areas, including countries bordering the Mediterranean. It is a clump-forming, succulent perennial. Unlike some other *Aloe* species, it does not form obvious stems or a low trunk, but instead spreads outwards, increasing by side shoots. The thickly succulent leaves are lance-shaped and pointed; on mature plants to around 70cm long, broadest at the base of the plant and slightly serrated along the margins. They are grey-green in colour; some plants feature white spots on the foliage. The leaves are rather brittle, especially on plants grown indoors and if broken readily ooze green-tinged, gel-like sap that is refreshingly cool to touch – it is this that is harvested commercially, mainly for the cosmetics industry. Plants growing in really hot, sunny places are sometimes rather yellowish or even red-tinged. Large plants will flower – the blooms look very much like those of *Kniphofia* (red hot poker). The individual flowers are yellow and tubular, nodding in impressive, tall, candle-like heads atop sturdy stems to 1m or more, which often branch and bear further heads.

In almost frost-free gardens, *Aloe* grows well beside other succulents in free-draining, even rocky soil in full sun. It looks great as a specimen clump in arid plantings growing through gravel, making a handsome garden or landscape plant and resisting drought easily – the occasional brief dip to around freezing may damage plants, but they soon recover as long as they are dry. Snowfall, however, causes more widespread injury. In temperate regions, *Aloe vera* are usually grown as house or conservatory plants and are easy to keep on a sunny windowsill,

Opposite Barbados aloe (*Aloe vera*) from *Flora Graeca* by John Sibthrop and James Edward Smith, 1823.

3
4
7
6
Aloe vulgaris

Opposite Barbados aloe (*Aloe vera*) from *Plantae Medicinales* by Theodor Friedrich Ludwig Nees von Esenbeck, 1828-33.

suffering few pests or diseases. They like free-draining compost such as that used to pot up cacti, and are best grown in a wide, weighty, terracotta container because plants can quickly get top-heavy and will soon develop offsets to form a sizeable clump. These offsets are easily detached with a little root and potted up – doing this may encourage the mother plant to grow larger and more impressive. If the original gets too big, you can simply discard and start again with a youngster.

This plant can make a handsome patio plant for standing outdoors with other tender succulents such as *Agave* and *Aeonium* in summer; plants revel in hot, sunny conditions and its drought tolerance means it needs little watering. When bringing plants outside they will need acclimatizing to the sun – stand them in semi-shade initially, or you risk burning foliage. There are other closely related plants that may prove hardier outdoors year-round in sheltered temperate gardens.

Note: This plant contains aloin which is toxic – no part of the plant should be consumed unprocessed. Sap may be applied topically; some people use it on burns or scalds. Do this with care as allergic reactions are widespread.

OTHER SUCCULENTS TO TRY

- ***Aloe arborescens*** is a handsome tall-growing species for a big container in a conservatory, but can be stood outdoors for summer. It is tall-growing with rosettes of serrated succulent foliage held on stout stems with spikes of showy red flowers. Keep frost-free.
- ***Aristaloe aristata*** is one of the hardiest. Low-growing, it forms tight rosettes of silver-marked dark green leaves and spreads slowly. Slender stems of coral pink flowers arise in summer. Good choice for a sunny rock garden with sharp drainage. Hardy to -5°C.
- ***A. ferox*** is a large stocky plant, forming huge rosettes of bold, fleshy toothed lance-shaped leaves held on a sturdy stem. It bears spectacular branched candles of red flowers. A good conservatory plant that may be stood outside in summer. Keep frost-free.

Aloiampelos striatula is in a genus closely related to ***Aloe***, and despite its exotic looks proves fairly hardy in mild gardens. Sprawling stems bearing fleshy, dark green leaves are topped by candles of yellow and green flowers. Needs a dry sunny place, hardy to -5°C.

Mexican lily

Beschorneria yuccoides

This remarkable and highly exotic-looking plant teams handsome foliage with a spectacular display of flowers. It is a perennial for a sunny, sheltered site, and in temperate gardens is best with the protection of a warm wall.

Opposite Mexican lily (*Beschorneria yuccoides*) from *Curtis's Botanical Magazine*, 1860.

Native to Mexico but naturalized in parts of Argentina and New Zealand, *Beschorneria yuccoides* is a large clump-forming perennial, with, as its name suggests, more than a passing resemblance to some species of *Yucca*, to which it is related. It forms similarly impressive rosettes of silvery or blue-green, pointed lance-shaped leaves to around 60cm long, but the foliage is much friendlier than that of its near cousin; quite soft and spine-free and although rather fleshy, not really succulent like those of *Agave*, also near relatives. In the wild, *Beschorneria* grow on dry cliffs and the sheer sides of ravines and so experience just about the sharpest drainage imaginable and prolonged spells of drought. There are several different species, but *B. yuccoides* is the one most frequently met in gardens; when grown well it makes a truly memorable sight, particularly when in flower. As with some yuccas and most agaves, individual rosettes are monocarpic – that is, they die after flowering – but happily this plant is relatively quick-growing and generous in its production of offsets. By the time a plant is ready to bloom, chances are a decent clump will already have formed.

The flowering spike emerges from the centre of the rosette, usually in early summer, occasionally running the risk of damage from late frosts in temperate gardens – gardeners are well advised to stand by to protect with fleece should a cold night threaten. The stem of the flowering spike is an impressive contrasting brick red colour and initially crook-necked and featuring showy, flamingo-pink, leaf-like bracts. As it develops, the bracts wither, those at the terminal end shrivelling to reveal masses of dangling red, green-tipped tubular flowers on what is now a flower spike 2.5m long; an extraordinary sight. The spike usually has a rather drunken stance, growing out diagonally rather than upwards – which is less curious when you remember its wild cliffside home. Green seed pods, shaped like a rugby ball, may form after flowering, but usually it is best to remove the old flower stems and then extract the dead rosette of foliage once it has withered, allowing space for younger rosettes to take its place. When a clump is mature you may get several flowering stems in a single year; plants take around five years to bloom.

Opposite Mexican lily (*Beschorneria yuccoides*) from *Curtis's Botanical Magazine*, 1882.

This Mexican wonder is not especially hardy. Plants will survive -5°C, but much lower and damage can be severe – if the ground freezes, you may lose the plant. As a result, in temperate gardens, plants are often seen planted right up against sunny sheltered walls, where they more readily survive cold winters than plants out in the open. They like full sun but will stand a little shade during the hottest part of the day while long spells of dry weather are easily shrugged off. That said, they also seem to thrive in much wetter climates than you might expect – they are a triumph growing on the house terrace at Mount Stewart in Northern Ireland. Plants must, of course, have sharp drainage – they are good in large, raised beds or gritty or sandy soil and will enjoy growing on banks that mimic their wild home, but the site needs to be sheltered. In colder regions *Beschorneria* can be planted in a glasshouse border or even kept in a large container and moved out for summer.

Plants look best in the company of other drought-tolerant exotics; *Yucca*, hardier *Agave* and *Opuntia*, perhaps alongside bulbous plants such as *Watsonia* and *Crocosmia*, or contrasting with hardier South African *Protea* or Australian bottlebrushes. In wetter climates they might look good backed by hardy banana *Musa basjoo*, glaucous *Melianthus major* or generous clumps of leafy *Canna*.

OTHER SIMILAR PLANTS TO TRY

- ***Beschorneria septentrionalis*** is a smaller-growing but desirable species with apple-green leaves, forming rosettes perhaps 50cm across. Its flower stems are similar in form but to around 1.2m long and pillar-box red in hue. It's slightly more tender but easier to manage and will do well in a container. Hardy to around -3°C.
- ***B. yuccoides* 'Flamingo'** is a highly desirable variegated selection with generously gold striped foliage and the same pink flower spikes. Hardy to -5°C.
- ***Furcraea longaeva*** is a closely related plant, if anything even more extraordinary, bearing a tree-like flower spike to 5m above 2m rosettes, and producing hundreds of dangling greenish white flowers. Needs to be kept frost-free.

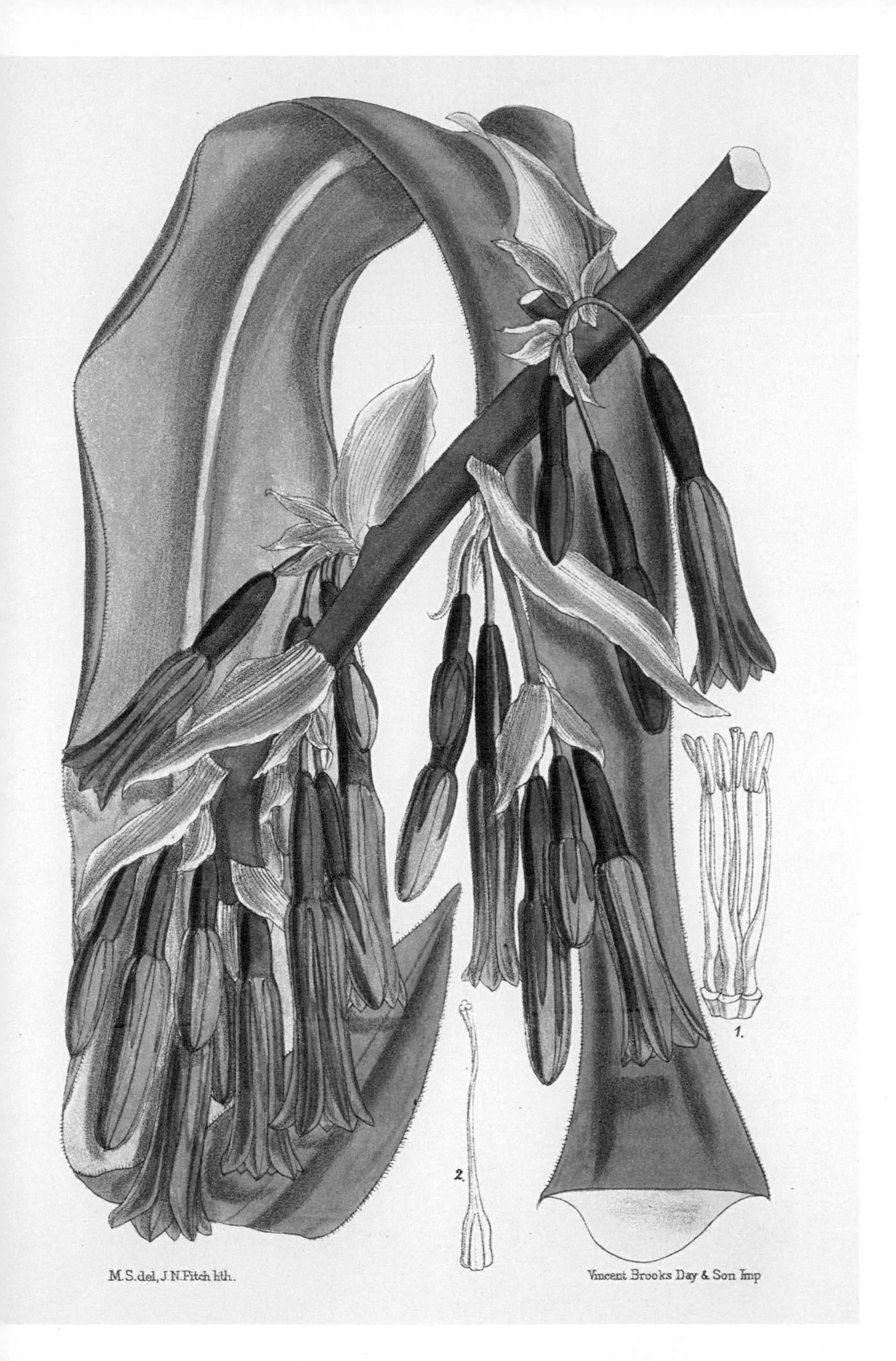
1.
2.
M.S. del, J.N. Fitch lith.
Vincent Brooks Day & Son Imp

Bougainvillea

Bougainvillea glabra

Few plants can compete with the vivid flower heads of this well-known tender climbing plant, originally from South America but now grown worldwide outside, or in greenhouses and conservatories – small examples can even make great summer patio plants.

Admired the world over for its dazzling displays of vibrantly hued, petal-like papery bracts, *Bougainvillea* are tender, climbing, usually evergreen plants grown outdoors in tropical and subtropical regions, and under glass where frosts are common. Heat- and drought-tolerant when established, they are a classic component of many Mediterranean, Caribbean and Californian gardens, and can often be seen scaling villa walls or cascading from old trees; long, questing stems are armed with backward-facing thorns which help plants climb. What usually pass for individual flowers are in fact large, showy, rather papery bracts (usually three, or six in double 'flowered' selections) – these encircle a cluster of three true flowers which are tubular and white clustered usually at the ends of branches but in some selections along their length. *Bougainvillea* are native to South America, found wild in Argentina, Brazil and into Peru and introduced to Europe in the early nineteenth century; soon they were grown in gardens around the world. There are around 18 different species but *B. glabra*, *B. spectabilis* and *B. peruviana* are most widely cultivated, alongside the many selections of *B.* × *buttiana*, the hybrid of *B. glabra* and *B. peruviana* available with large bracts in hues of white, pink, purple, red, orange and yellow. A wide range of different selections can be found at specialist suppliers and plants may bloom virtually year-round.

Probably still most widely grown, *Bougainvillea glabra* is purple-pink in flower and reaches around 8m tall, sometimes more. The oval, pointed leaves may be bronze-tinted when young; the individual flower heads are carried in branched clusters on the current year's growth, forming impressive trusses of bloom. It is often seen trained as a hedge and can be grown as a free-standing standard tree-like plant, rather in the style of some *Wisteria*; plants can even be kept as bonsai. Outdoors it needs an open place in full sun, in well-drained, slightly acid soil and will need wires or trellis initially if it is to climb. Once established, plants actually flower better if grown in dry conditions – too much water promotes leafy growth. In extreme drought, plants may defoliate but will re-sprout when rain arrives. Plants are usually pruned after flowering to keep them

Opposite
Bougainvillea (*Bougainvillea glabra*) from *Revue horticole*, 1889.

b.

g.

d.

f.

h.

a.

e.

BOUGAINVILLEA GLABRA
VAR. SANDERIANA.

H. Schade

Chromolith. Fr. Eugen Köhler, Gera-Untermhaus.

Opposite
Bougainvillea (*Bougainvillea glabra*) from *Gartenflora*, 1899.

within bounds, usually during winter (which in ideal conditions can be a small time-window). They make a great backdrop to displays of succulents and cacti, or tumbling from a pergola, helping to provide shade.

In temperate regions, *Bougainvillea* is too tender to keep outside all year, tolerating only light frost. Small plants can however be grown on sunny windowsills or in a greenhouse in containers. Make sure the pot has good drainage holes. This can be moved outside for summer, to a sunny, sheltered position, making an impressive focal point when in flower. Plants need feeding with nitrogen-rich fertilizer to promote growth, but when in flower, use tomato fertilizer to help prolong the displays. Larger plants are great planted in sunny conservatory or greenhouse borders, trained on wires up walls and along the roof, they can flower for several months. They like well-drained, fairly dry soil; in hot conditions provide plenty of ventilation and maintain humidity. Keep above 10°c in winter if you want plants to retain foliage. If cooler, at around 5°c, plants will lose leaves, re-shooting in spring – keep plants almost dry until then.

OTHER BOUGAINVILLEA TO TRY

- ***Bougainvillea* × *buttiana* 'Poulton's Special'** features masses of large, rich, almost shimmering pink bracts – an excellent selection.
- ***B.* × *buttiana* 'Raspberry Ice'** cerise pink bracts contrast with generously variegated foliage, the leaves broadly edged in cream. A good one for a pot, and said to tolerate lower temperatures well.
- ***B. glabra* 'Sanderiana'** is one of the best-known selections and said to be one of the most cold-tolerant, with impressive displays of pink-purple bracts.
- ***B.* 'California Gold'** produces flushes of rich golden-yellow bracts along the plant's branches, even into winter in warm seasons. It proves vigorous but needs a warm position.
- ***B.* 'Los Banos Beauty'** is a delicately hued selection which produces white bracts tipped in pale pink; a beautiful combination of colours.
- ***B.* 'Pedro'** features startling pillar-box red bracts – it is also a relatively compact selection so a good choice for growing in a container.
- ***B.* 'Sundown Orange'** is an especially brightly hued selection, with large clusters of bracts that change colour as they age, starting out rich orange and changing to coral pink, then soft pink before falling.

Chinese trumpet vine

Campsis grandiflora

This climbing plant for a warm house wall is sensational at the end of a sunny summer, producing clusters of large funnel-shaped, pinkish-orange and red flowers well into autumn.

Opposite Chinese trumpet vine (*Campsis grandiflora*) from *Flora Exotica* by Heinrich Gottlieb Ludwig Reichenbach, 1835.

No other hardy climbers produce floral displays that can compare with the end of summer spectacular provided by *Campsis*; masses of large trumpet- or funnel-shaped flowers in fiery tones of orange and red, held in impressive trusses high on stems well clear of the generous and rather handsome pinnate foliage. There are two species in this small genus: *C. grandiflora* which hails from China, arguably the more spectacular with its larger flowers, and *C. radicans* from south-eastern states of North America, which is also impressive and more often seen in cooler gardens because it tends to be more accommodating in its requirements. The hybrid between the two is C. × *tagliabuana* which provides gardeners with more fine plants to choose from. They are all vigorous self-clinging deciduous climbers, hardy to around -10°C and lovers of hot, sunny sites – highly drought-tolerant when established, and in the wild climbing up through trees, but in gardens favourites for scaling sunny house walls, well capable of reaching the first floor and growing over the roof if allowed. After flowering, finger-like seed pods may form in a hot summer, most often on *C. radicans* in cooler climates; this species also suckers from the root freely.

In warm continental and Mediterranean climates, *Campsis grandiflora* is widely grown, and was introduced to the West at the beginning of the nineteenth century. It makes a highly reliable choice for flowering at the end of summer, often bridging the seasons with its flamboyant flowers. The large funnel-shaped blooms have yellow throats enlivened with reddish markings, while the five generously flared and rounded petals glow in tones of pinkish-peach and orange. They are held in profuse terminal clusters weighing down the ends of every arching lateral branch, large plants providing a terrific show and attracting pollinating insects from afar. For any *Campsis,* choose the hottest, sunniest wall you can find – ideally a south-facing corner where the plant can really bake. This is most important for *C. grandiflora*, which seems to need more summer heat to ripen stems and produce flowers than the others, although it is probably of similar winter hardiness. Soil is less important as long as it is well drained – best results often come from rather poor, dry stony sites.

Opposite Chinese trumpet vine (*Campsis grandiflora*) from General John Eyre Collection, c. 1800s. Kew Collection.

Campsis climb via adventitious roots produced from their long stems, which makes them perfectly adapted to use for covering large walls. Trellis and wires are not required in the long term and are unlikely to support the growth these vigorous plants can quickly make once established; they will soon swamp an arch, fence panels or small pergola, for example. Young plants can take a bit of time to attach themselves to whatever they are required to climb, especially in windy sites – wires or trellis will be needed at this stage. Flowering can take a few years: they'll often start to bloom once they have reached the top of their support, the growth habit changing from climbing to flower-bearing; the higher the wall, the longer flowering is likely to take, but when happy, growth is rapid. Mature plants may threaten to outgrow their space; an annual prune in spring to cut back summer growth is sensible, and keep an eye out for still climbing stems that threaten to probe below roof tiles, for example.

In southern Europe it is common to see *C. grandiflora* growing on quite low, free-standing boundary walls, perhaps 2m tall, the plant cascading over and producing arching flowering shoots 1–2m long, perhaps in the company of vines, *Bougainvillea* and *Plumbago*. With ever-warmer summers, this way of growing is becoming more possible in more temperate climates too, although more appropriate partners include clematis and climbing roses.

OTHER *CAMPSIS* TO TRY

- ***Campsis radicans*** is a strong growing climber, requiring a sunny wall to thrive but often easier to please than ***C. grandiflora***, its fiery red and orange trumpets with less flared petals. The selection **'Atrosanguinea'** has darker red blooms, while lovely **'Flava'** bears soft yellow flowers each with a rich orange-yellow throat.
- ***C. × tagliabuana* 'Indian Summer'** (or **'Kudian'**) is a recent rather distinctive introduction and rather less vigorous, producing large, soft orange trumpets, each with a deep red throat.
- The most popular hybrid ***campsis*** in cooler, temperate climes is probably long-grown **'Madame Galen'** with orange flowers that have a pinkish tint. It's a vigorous and reliable free-flowering choice for sun.

53.

HONG KONG, GENERAL J. EYRE. delt.
Purchased, 1904.

Californian lilac

Ceanothus arboreus

Loved by gardeners for its impressive panicles of blue flowers, bold leaves and helpfully quick-growing habit, this sun-loving evergreen makes a useful shrub for dry sites.

Among the most reliable of hardy plants for dry gardens are *Ceanothus,* quick-growing evergreen or deciduous shrubs native mostly to arid western states of North America, particularly California, renowned for their blue (sometimes white or pink) flowers produced in such dazzling profusion that they may smother plants for weeks, usually in early summer, sometimes autumn. Some are low-growing, others upright or bushy, a few are tree-like. The largest is *C. arboreus*, known only from the tiny eight Channel Islands of California and one site in Mexico, where it is found in scrubby coastal plant communities and chaparral. Plants form as slightly sprawling shrubs or trees, growing rapidly to 8m (it will reach 6m in five years) developing a wide canopy of rich green, glossy, oval leaves 10cm long with silvery undersides. The fragrant blue flowers are carried in upright 12cm pyramidal panicles that look almost like those of lilac and attract pollinating insects.

Opposite Californian lilac (*Ceanothus arboreus*) illustrated by Malcolm English, 2024.

This evergreen makes a great, quick-growing plant for a sunny place, away from cold winds, and sheltered by a warm wall, although it may get too large and vigorous for wall training. Better to let it develop a tree-like form; it needs space as it is apt to be wide-spreading, losing lower branches as it grows. Light pruning can improve its overall shape; this is best done right after flowering, but never to cut into wood thicker than a pencil. Young plants are best supported by a post because they get top-heavy; winds, heavy rain and snowfall can snap brittle stems. Although a sun lover, *C. arboreus* will do well with light shade for part of the day. It is at its best in late spring and early summer when flowering can be profuse, but often a second less generous flush of flowers appears in autumn. It makes a great choice for the back of a drought-tolerant shrub planting, alongside *Cistus*, lavenders, *Lupinus arboreus* and shrubby umbel *Bupleurum fruticosum*.

Generally, *Ceanothus* prove drought-resistant but young plants do need water in the first year. Some of the hardiest survive -10°C; *C. arboreus* is not quite as tough, standing perhaps -7°C. All *Ceanothus* revel in poor, sandy soil but may also survive on clay as long as it is not waterlogged.

Fountain grass

Cenchrus longisetus

This sun-loving ornamental grass, at its superb best late in the season, is known for its delightfully fluffy flower and seedheads produced in multitudes above mounds of arching foliage.

Opposite Fountain grass (*Cenchrus longisetus*) illustrated by Malcolm English, 2024.

Loved by many, especially younger gardeners for their late summer and autumn displays of irresistibly tactile, fluffy flower and seedheads, the grasses of the *Cenchrus* genus (previously known as *Pennisetum*) are highly valued for their ornamental attributes in the garden, and are known to stand spells of heat and drought well, once established. Some are reasonably reliable perennials in temperate gardens, others distinctly tender and best considered annuals for late season appeal in borders or containers. From Arabia and parts of seasonally dry tropical eastern Africa comes *Cenchrus longisetus* (syn. *Pennisetum villosum*), a mound-forming but spreading grass with slender, rather rough to the touch, arching, mid green foliage. At flowering time, this is overtopped by masses of delightfully fluffy, feathery heads, up to 10cm long and 2cm wide, which are initially green in hue but change over several weeks through parchment brown to almost white, the plant reaching perhaps 90cm tall and as much or more across.

Given the plant's natural range, this ornamental grass may not sound especially promising as a reliable perennial in gardens where frost is common; it has been cultivated since the nineteenth century but is increasingly seen in our warming climate because it now often proves enduring as well as endearing. In frost-free climates, however, it does need watching closely because plants can become something of an invasive pest; the species has naturalized in many countries, including Australia and North American states, such as California, Texas and Hawaii, spreading by seed and underground rhizomes.

In temperate gardens, *Cenchrus longisetus* likes a sunny, open site in any well-drained soil. It is a perfect candidate for injecting end-of-season interest into borders, its pale flower heads at their most showy in autumn, standing out beautifully from the glowing colours of falling leaves elsewhere in the garden. Avoid planting too densely because individual clumps have a naturally handsome arching form – this is a good plant for a gravel garden where plants are well spaced. It also suits the edges of borders where hands and legs will brush by, and looks highly effective planted in a drift with other grasses and late interest perennials, such as *Kniphofia*, *Verbena bonariensis*

Opposite *Cenchrus orientalis* from the Mary Maitland Collection, 1904. Kew Collection.

and *Rudbeckia*. Its low stature is also useful because it never needs support and allows views to planting beyond. In coastal gardens it succeeds well, thriving on and helping to stabilize banks with its underground runners. When winter starts to bite, the plant will begin to collapse and loose its foliage – it stands temperatures down to -5°C, perhaps lower if a mulch of garden compost is applied over its crown. Do not despair if there is no sign of new growth in early spring: it is often slow to begin into growth. Once new shoots are spied, usually in May, trim away the remains of last year's foliage to tidy.

OTHER ORNAMENTAL GRASSES TO TRY

- ***Ampelodesmos mauritanicus*** (vine grass) is a spectacular architectural evergreen grass native to dry Mediterranean regions, reaching 2m with tall upright stems bearing drooping heads of flowers that look great well into autumn. Likes sun. Hardy to -5°C, possibly lower.
- ***Anemanthele lessoniana*** (pheasant's tail grass) provides colourful evergreen appeal with its billowing clumps, 60cm tall, of orange-tinted, olive-green blades, which in summer are softened by shimmering clouds of minute flowers. Likes sun or part shade. Hardy to -10°C.
- ***Cenchrus alopecuroides*** (syn. ***Pennisetum alopecuroides***) is a relative of ***C. longisetus,*** producing larger, more spectacular fluffy flower spikes almost like bottle brushes 15cm long, standing erect on stems to 1m or more, above arching foliage; named selections are available, of differing stature and flower tint. Likes sun, hardy to around -5°C.
- ***Festuca glauca*** (blue fescue) makes a great choice for a small garden; it forms small clumps of bright blue foliage to around 30cm tall and as much across. In early summer small flower spikes form. Needs sunny, well-drained place. Hardy to -15°C.
- ***Helictotrichon sempervirens*** (blue oat grass) proves highly ornamental with its blades of blue green forming an elegant fountain of foliage to 60cm, overtopped by sprays of green oat-like flowers. Full sun, good drainage, hardy to -10°C.
- ***Celtica gigantea*** (giant oat grass) makes a magnificent sight in flower with tall stems to 2m bearing long-lasting, golden oat-like flowers that remain in beauty after the seeds have fallen, well into autumn. Full sun, hardy to -10°C.
- ***Cenchrus orientalis*** (syn. ***Pennisetum orientale***) is another great species with pink-tinged flower plumes in autumn to 70cm. Hardy to -5c.

Painted by MRS. GEORGE GOVAN (Mary Maitland), circa 1823–1832. Presented by her sons, DOUGLAS M. GOVAN, Esq., and Major-General C. M. GOVAN, R.A., 1904.

Pennisetum orientale, Rich.
Var. triflorum

मूकामसिमन्लामाहसितंबरनामघोटणा

Gothnu

Ghoṭaṇā. Simla – Sep.

Judas tree

Cercis siliquastrum

This compact, hardy, flowering tree is resplendent in spring when masses of small, pea-like flowers garland its branches, then in summer its distinctive heart-shaped leaves form a broad-spreading canopy, providing cooling shade below.

Opposite Judas tree (*Cercis siliquastrum*) from *Curtis's Botanical Magazine*, 1808.

This beautiful and useful small deciduous tree sadly seems forever linked with the death of Judas Iscariot, for it is often said that the disciple hanged himself from its branches after betraying Jesus Christ and that this action caused its previously white flowers to turn red (in fact, the flowers today are usually rich pink). Certainly, this species is native to the Holy Land, as well as other parts of western Asia and southern Europe, and often found in dry, rocky places, a habit and distribution that hint firmly at drought- and heat-tolerance. More generally, the genus *Cercis* is surprisingly widespread, growing in warm, temperate parts of the world, with species native to North America and Mexico and eastern Asia, across into China; they are collectively known as redbuds and of similar stature to *C. siliquastrum*; in recent years many named cultivars of these have been selected, as gardeners begin to appreciate their attractive, yet easy-to-please characters.

Long and widely cultivated in the Mediterranean, *Cercis siliquastrum* is usually a multi-stemmed tree, reaching around 10–12m tall at most, but broad-spreading, forming a dome-shaped canopy in time. Its leafy summer crown is often put to good use as a shade tree in parks and gardens throughout the region, but it is in late spring when this tree most impresses, producing masses of those small but strident pink flowers. Attractively these develop and open before foliage, wreathing bare stems and branches in a most eye-catching show. They emerge not only from tree branches in the usual way but, more curiously, directly from the bark of mature branches – even the trunk – in a habit known as cauliflory, more often seen with some tropical trees. Close inspection of these flowers reveals them to be shaped like members of the pea family; a relationship confirmed after flowers are followed by conspicuous crops of flattened, purple-tinged seed pods that dangle from branches, ripening through summer. In some countries the flowers of *Cercis* are eaten – often in salads – and are said to have a rather bitter flavour. The foliage of these trees is highly appealing too: rich green individual leaves rounded and more or less heart-shaped. Foliage interest is an attribute pronounced in some cultivars of species such as *C. canadensis* (see below).

CERCIS SILIQUASTRUM

Opposite Judas tree (*Cercis siliquastrum*) from *The Garden*, 1892.

In gardens and particularly urban settings, even as street trees, *Cercis* are proving increasingly valuable. They revel in sun and heat, and once established seem able to shrug off drought conditions, yet they also prove hardy, *C. siliquastrum* standing -15°C.

In cooler gardens they are occasionally seen grown as striking wall-trained specimens because summer heat is needed for good displays of flower. They are useful too as specimen trees, making impressive focal points in gravel gardens, for instance; or grouped, perhaps used formally as a pair to frame a view; or even as an avenue along a drive. Their relatively compact size means they fit well into many smaller sites too; multi-stemmed plants have a more natural appeal, but standard trees trained up on a single trunk are also available.

Choose a well-drained, open position and secure young trees with a tree stake, watering regularly in the first year until established. The plant will associate well with other sun-loving plants from arid lands; try beside *Phomis fruticosa*, *Cistus*, *Salvia*, *Genista*, lavenders and *Nepeta*; the broad rounded crown in time contrasts beautifully with the green pillar-like form of *Cupressus sempervirens*. As the crown begins to cast more shade, introduce plants such as *Cyclamen hederifolium* or spring-flowering *C. coum*, *Euphorbia*, *Helleborus* and *Geranium* below.

OTHER *CERCIS* TO TRY

- ***Cercis siliquastrum* 'Bodnant'** is a particularly fine garden selection with rich purple-pink flowers; clean white-flowered selections are also available, sold as ***C. siliquastrum* f. *albida***; they make a great contrast with the more usual pink. Hardy to -15°C.
- ***C. canadensis* 'Forest Pansy'** is perhaps the best known of the coloured leaf redbuds with superb large rich red-purple heart-shaped leaves; it also bears pink flowers when mature and stands some shade and shelter. Marvellous **'Ruby Falls'** is a weeping selection with red leaves, while impressive **'Hearts of Gold'** has bright yellow foliage. Hardy to -15°C.
- ***C. chinensis*** is similar to ***C. canadensis*** but with rather glossier leaves; the free-flowering selection **'Avondale'** is often sold bearing rich purple blooms. Needs full sun; hardy to -15°C.

Peruvian apple cactus

Cereus repandus

This spectacular, architectural, easy-to-grow cactus thrives in dry gardens that experience little or no frost; in colder areas, young plants are impressive grown indoors during winter and moved out on to the patio through the warmer months.

Arguably the tallest of all cacti, this mighty tree-like plant is a true wonder of the natural world, producing remarkable cylindrical, succulent stems that can easily soar to 10m. Given support, these have even been recorded exceeding 30m, higher than any other cactus. Also commonly known as kadushi, this plant was originally native to arid regions of South America and the Caribbean islands, but has long been widely cultivated, both for its ornamental appearance and the edible fruit it bears, in dry tropical or subtropical countries around the world; in South Africa it is considered a weed.

Its erect, segmented stems are distinctive and very appealing – rich blue-green in hue, even as a young plant – but when mature, plants often develop a single trunk-like main stem, eventually with the thickness of a man's leg. This bears arching, then upright growing side branches, usually to 10cm thick, with around 10 shallow ribs, which are furnished with clusters of spines up to 5cm long. The summer flowers, 15cm long, are spectacular and produced in flushes. Each bloom opens from scaly buds that emerge from the stems, and open for one night only. They are white and funnel-shaped, with pink- or green-tinged outer sepals and produce a sweet fragrance to attract night-flying pollinators. This cactus usually requires cross-pollination to form fruit, but when this occurs they can be produced quite freely with hundreds appearing on mature plants – they are around 10cm across and roughly apple-shaped, with an inedible leathery red-purple skin. Inside the edible, crisp white flesh is spotted with crunchy black seeds – it has a sweet flavour and plants are grown commercially for their fruit in some areas; unlike prickly pears, the fruits are thankfully spine-free.

Perfectly adapted to surviving drought conditions, this is an easy, quick-growing cactus, putting on 30cm a year when happy. It is sometimes seen planted as a hedge because individual stems can be rooted as cuttings when quite large. It looks great in arid garden plantings alongside other succulents, including *Euphorbia*, *Agave*, *Aloe* and *Furcraea*, or other cacti – they contrast well with *Opuntia* – or palms such as *Chamaerops* or *Washingtonia*.

Opposite Peruvian apple cactus (*Cereus repandus*) from *Revue de la Famille des Cactées* by Augustin Pyramus de Candolle, 1829.

Opposite Peruvian apple cactus (*Cereus repandus*) from *Revue de la famille des cactées* by Augustin Pyramus de Candolle, 1829.

Choose a position ideally in full sun – although it will stand a little shade for part of the day – in a place with sharply drained soil. This cactus will tolerate short frosts, surviving temperatures of -5°C for short periods if kept dry at the root, which means it can also succeed planted outdoors in Mediterranean climates; the protection of wall will help here. In cooler, temperate countries where frosts are common, it is easily grown as a houseplant or cultivated under glass – in hot summers it even makes an unusual patio plant moved outside for the season and planted in a large container. Take care when positioning initially because bright sun can burn plants grown indoors. Although obviously drought-tolerant, as with other cacti, water through the summer promotes good growth. During autumn, before hard frost and heavy rain, bring plants into a greenhouse or conservatory and keep them almost completely dry over winter, and ideally above 5°C. Young potted plants usually grown from cuttings make good houseplants for a sunny windowsill; if repotted regularly and kept in enough light, they will grow quickly and can soon become impressive specimens.

OTHER COLUMNAR CACTI TO TRY

- ***Cereus aethiops***, also from South America, is a smaller-growing cactus reaching around 2m, again with blue-green stems, although they are usually rather prickly, the spines dark in colour. It flowers well as a young plant. Possibly hardier if dry, to around -7°C.
- ***C. forbesii* 'Spiralis'** is a popular cactus grown for the fascinating spiral growth pattern of its stems – it also bears large white flowers. Good for a windowsill when young; keep frost-free.
- ***C. repandus* 'Monstrosus'** is a selection of Peruvian apple cactus with a curious irregular growth pattern, looking almost like a living sculpture. Survives brief spells down to -5°C if dry.
- ***Lophocereus marginatus***, a handsome so-called organ pipe cactus from Mexico, eventually makes a tall plant with individual stems remaining unbranched. Best kept frost-free.
- ***Stenocereus griseus*** is a tree-like cactus from South America, where it is often planted as a hedge – young plants are good on a windowsill and stood out for summer. Keep frost-free.

Dwarf fan palm

Chamaerops humilis

Compact and easy to grow, especially in hot, sunny locations, this attractive palm makes a great choice for adding an exotic feel to plantings and is able to survive weeks of summer drought as well as considerable winter cold.

Opposite Dwarf fan palm (*Chamaerops humilis*) from *Phytanthoza Iconographia* by Johann Wilhelm Weinmann, 1745.

Few species of palm are really reliable for year-round effect in cool, temperate gardens, but this sun-loving European native can usually be counted on, except in the very coldest winters; it is also a great plant for surviving long periods of summer drought, seemingly without the slightest worry. The only species within the genus *Chamaerops*, *C. humilis* is found all around the Mediterranean, thriving often in poor, rocky ground and forming a part of coastal maquis habitats. It is a clump-forming plant with tightly clustered shoots originating from a single base. Plants are slow-growing and only gradually develop rather stout rough trunks – more often they are seen forming a dome-shaped leafy mound around 1.5m high – but in time old mature plants may build into impressive multi-trunked specimens to around 6m. Occasional single-trunked plants are encountered – usually in gardens where side shoots have been removed to promote a palm tree-like form. The large, rounded fan-shaped fronds can reach around 70cm long, held on a long leaf stalk with a sharply toothed base. The leaf itself has up to 20 slender finger-like divisions and is usually green with conspicuous silvery undersides, although some plants of *C. humilis* var. *argentea* from the Atlas Mountains have striking glaucous leaves. Spikes of yellow flowers appear in summer – plants are dioecious which means they have male or female flowers; pollination (which is carried out at least in part by a particular kind of weevil) results in bunches of small rounded reddish fruit that can look rather like grapes.

Cultivated since the early nineteenth century and now seen in gardens around the world, this palm is today proving a useful ornamental plant as the climate warms and rainfall becomes more unpredictable. It is widely planted as an ornamental, even in its native habitat and often seen as an amenity species, growing in towns and cities in southern Europe. Happily, it seems reasonably resistant to deprivations of introduced palm weevil (*Rhynchophorus ferrugineus*) which has devastated ornamental palm species such as *Phoenix canariensis* (Canary Island date palm) in many areas, although it is known to act as a host.

Opposite Dwarf fan palm (*Chamaerops humilis*) from *Botanist's Repository for New and Rare Plants* by Henry Charles Andrews, 1809.

Of rather similar appearance is close relative *Trachycarpus fortunei* (Chusan palm), an Asian species and perhaps the best-known palm of all for frost-prone climates and one that does well with some shade, wind shelter and, importantly, moisture. *Chamaerops* will succeed far better, however, in those open places that are baked by summer drought and heat. It thrives even in full coastal exposure, a location that soon reduces larger-leaved *Trachycarpus* to tatters. Although regarded as less hardy, *Chamaerops* usually stands -10°C; in very cold winters some additional protection may be needed to get it through.

Dwarf fan palm does best in well-drained soil in full sun – reflected heat from a wall can be beneficial in particularly cold districts; it will stand some shade for part of the day, however.

You could consider growing the plant in a large pot; this palm adapts well to life in a container and can be spectacular. It also looks impressive in gravel gardens, exotic borders and even large rock gardens, bringing a feel of the tropics to plantings and forming an arresting living focal point once well established. You can combine it with succulents, potted cacti and *Cycas* for a really arid impression, or use them to provide contrast with Mediterranean *Cistus*, rosemary, lavender or *Artemisia*. Otherwise, if water use allows, display them beside hardy banana *Musa bajoo* and bedded out cannas for a full-on tropical effect.

OTHER FAN PALMS TO TRY

- ***Chamaerops humilis* 'Vulcano'** is a curious selection of dwarf fan palm from Sicily – it is very compact, with stiff, rounded, rather silvery leaves held on short stalks. Hardy to -10°C.
- ***Trithrinax campestris*** (Caranday palm) from Argentina is a ferocious customer that resembles ***Chamaerops*** closely, with stiff, silver spine-tipped leaves that top trunks clad with needle-like spines. Impressive from afar and survives drought and flooding. Hardy to -10°C.
- ***Washingtonia filifera*** (desert fan palm), from south-western USA and Mexico, is an impressive large-leaved fan palm growing to 18m. The large leaves display filament-like threads. Best in a Mediterranean climate. Mature trees may survive -7°C.

Sun rose

Cistus × purpureus

This easily grown, sun-loving evergreen shrub makes a marvellous show in early summer when the plant is smothered in masses of large, fragile, distinctively spotted purple flowers for weeks on end.

Perfectly suited to dry, sunny, well-drained places, *Cistus* are valuable shrubs for any garden where drought is a common threat. They are all evergreen, the foliage of some resinous and aromatic, and they bear a succession of delicate, short-lived five-petalled flowers often in pink, but also white or purple – a few also display contrasting blotches at the base of each petal. Most are bushy and reach around 1m tall (a few are rather larger) but often sprawl to be wider than they are tall. There are around 20 species in all, native to the Mediterranean region, including the Canary Islands, and they thrive in poor, rocky ground; many plants in gardens are hybrids or named selections with ornamental attributes. They are closely related to *Helianthemum* and *Halimium*, which enjoy similar growing conditions and are also excellent garden plants.

Popular in cultivation, *Cistus* × *purpureus* is a large flowered hybrid between *C. creticus* and *C. ladanifer* and has been known since the late eighteenth century. It forms a wide-spreading, broadly domed plant with reddish stems and long lance-shaped, dark green leaves. The floral display begins in late spring and lasts around a month; masses of flowers, 8cm wide, are produced from the tops of new shoots, each fragile bloom with shimmering rich purple, rather creased petals that look almost like crumpled satin. At the base of each petal is a dark red spot. The striking selection 'Alan Fradd' is similar but has white petals, the odd petal marked with purple. It is a fast-growing plant and, like most other *Cistus*, quickly rewarding in the garden, filling space and providing impressive displays.

Sun roses revel in sun but will stand a little shade through the day. Once established, plants stand summer drought well, but need some extra water initially after planting – a job best done in spring. Well-drained soil is ideal – they thrive especially well on poor or chalky soil but will often tolerate heavier ground as long as it does not get too wet in winter. Shelter from cold is useful; in severe winters plants can be damaged or even killed – most are hardy to around -10°C and a few, such as white-flowered *C. laurifolius* or popular *C.* × *hybridus*, possibly lower. In chilly districts, grow them in a border backed by a warm wall.

Opposite Sun rose (*Cistus* × *purpureus*) from *Cistineae* by Robert Sweet, 1825–30.

Opposite Sun rose (*Cistus × purpureus*) from *The Garden*, 1887.

These plants look superb tumbling down a sunny bank or spilling over the edges of retaining walls in raised beds, or relaxing their branches over the edges of a sunny patio; they are also great plants for maritime exposure. All are perfect plants for adding structure and colour to gravel gardens, or for a Mediterranean-inspired terrace or border, associating perfectly with other drought-tolerant shrubs such as *Ceanothus*, *Phlomis fruticosa*, lavenders and rosemary and growing well in sun below taller *Acacia dealbata*, *Laurus* and *Cupressus* where shade is minimal. Contrast their relaxed sprawling forms with upright clumps of *Phormium* or *Miscanthus* grasses for drama.

As *Cistus* grow quickly and have a spreading instinct, there often comes a time when plants exceed their allotted space. These are not good plants for growing in containers long term and they won't stand being moved. It is often said *Cistus* cannot be pruned; while it is true that they will not tolerate hard pruning back into woody growth, many will stand light trimming after flowering, which helps to restrict plant size and prolong the garden life of these evergreens. That said, they are not generally long-lived and are usually best replaced after 8–10 years, if not sooner. Happily, most are easily propagated from cuttings taken in autumn.

OTHER *CISTUS* TO TRY

- ***Cistus × aguilarii*** is a handsome and vigorous, rather upright hybrid with large and arresting pure white flowers that stand out well from the dark green foliage. Reach 2m. Hardy to -10°C.
- ***C. × argenteus* 'Silver Pink'** is a highly popular, easily grown hybrid with large silvery green leaves and masses of peach-pink flowers. Hardy to -10°C.
- ***C. creticus*** forms a compact, shrubby plant seldom reaching 1m high with masses of flowers, bright pink and 5cm wide, and rather sticky foliage. Hardy to -10°C.
- ***C. × hybridus*** (syn. ***C. × corbariensis***) makes an excellent choice – it is compact-growing, tough and has dark green leaves and masses of flowers, pure white and 4cm wide, through early summer. Hardy to -10°C.
- ***C. × incanus*** (syn. ***C. × pulverulentus***) **'Sunset'** has soft green foliage and impressive displays of vivid magenta flowers. Hardy to -10°C.

CISTUS PURPUREUS.

Pampas grass

Cortaderia selloana

Well known in gardens around the world for its lofty silver or pink plumes, this impressive evergreen grass makes a tough drought-tolerant choice. While many selections need plenty of room, a few can even suit small spaces.

Opposite Pampas grass (*Cortaderia selloana*) from *Flore des serres et des jardin de l'Europe*, 1874.

This statuesque evergreen grass is a well-known ornamental in many cool, temperate gardens, where it is much admired for its upstanding flower plumes, which are produced at the end of summer and often endure on plants through to spring, in some plants to around 4m tall. It forms mounds of fountain-like growth, the narrow linear green-grey foliage to around 2m long, arching with considerable elegance, but armed with saw-edged, razor-sharp margins that can easily cut unwary fingers. It is native to South America, notably the Argentinian pampas, an open prairie-like grassland habitat where high summer temperatures are frequent, winters generally mild but with some frost, and rainfall sporadic in summer with long dry spells punctuated by heavy downpours. As a result, adaptable *Cortaderia selloana* can prove highly drought-tolerant and suited to growing in many regions; in countries such as Hawaii and New Zealand (which has its own elegant but, sadly, moisture-loving species) the plant is an invasive pest, spreading by seed, especially on disturbed land, and reducing biodiversity.

It has long been cultivated; introduced to the UK in the nineteenth century, it then spread to gardens around the globe. It was highly popular in Victorian times and found to be useful in large gardens and parks, where it can form an arresting landscape feature, often positioned beside water and with a contrasting backdrop for best effect. The flower plumes of *Cortaderia selloana* vary in colour – usually they are off-white but in some selections they can be silvery-white or even pink-tinted. The appearance of the flower plumes varies depending on the sex of the plant – female plants have more upright, rather silvery plumes, while male flowers tend to be of arching habit and off-white or pink. Numerous selections have been made; some are dwarfs suitable for small spaces while others have improved flowering or variegated foliage. *Cortaderia selloana* tends to be a very hardy plant, surviving -20°C and ideal in exposed, windy sites such as seafront-facing locations.

In gardens, plants prove tough customers. They do need full sun and an open position; plants will not tolerate waterlogging, and although they don't mind fairly moist conditions, they stand drought impressively once established.

Gynerium argenteum N. ab. Es.

Brésil (Pleine-terre.)

Opposite Pampas grass (*Cortaderia selloana*) from *L'Illustration horticole*, 1855.

Plant this grass somewhere you can appreciate the overall form of the clump, but away from paths where its leaves may cause injury. It can look terrific at the back of a border, holding its own with shrubs and small trees, or with large perennials – perhaps lofty *Eupatorium*, *Helianthus salicifolius* or *Rudbeckia laciniata* 'Herbstsonne'; or other bold grasses such as *Miscanthus* and *Arundo donax*. Autumn is when *Cortaderia* shines and the fiery tints from leaves and fruit of plants such as *Rhus*, *Acer*, *Malus*, larger deciduous *Euonymus* and *Cotoneaster* are particularly appropriate beside the fluffy plumes for a spectacular finale to the season.

Cortaderia maintenance is not to be taken lightly. Plants sprout from a woody crown, but as old foliage dies it is retained, building into a living haystack. The best action is to shear back plants with hedge trimmers (and sturdy gloves) in spring and pull out the worst debris. A traditional alternative is to set fire to clumps (supposedly to mimic occasional wildfires of the pampas). It can save time and effort if you take precautions (have a hose on standby) and don't mind a charred mound until the plant recovers. Make sure it is not close to wooden fences and don't set light to plants that have been neglected longer than a couple of years, as the fire may be so intense it kills them.

OTHER *CORTADERIA* TO TRY

- ***Cortaderia selloana* 'Aureolineata'** is one of the best selections, with leaves edged in gold and silver flower plumes to 2m. **'Pink Feather'** is one to choose for colourful plumes – they are rich pink and soar to more than 3m tall. By contrast **'Pumilla'**, is a distinctive compact plant with creamy-silver plumes to 1.2m. **'Sunningdale Silver'** is another giant to 3m while **Tiny Pampa ('Day1')** is one of the smallest, to just 80cm when in flower.

Italian cypress

Cupressus sempervirens

This much-loved conifer adds special character to the landscape in many warm, dry lands, forming narrow, pointed columns of green. In cooler climes it is also proving successful, and is a great way of adding height to planting without taking up much space.

With its slender, pillar-like form, this characterful conifer is so closely associated with southern Europe and the Mediterranean, the mere sight of it brings to mind Umbrian and Tuscan landscapes where this tree adds so much to the surroundings, hence its common name. The distinctive columnar shape makes it perfect for contrasting with more rounded trees and shrubs, while its fastigiate habit means it can bring considerable height to even small spaces, making it useful in many gardens. It is often seen planted as an element in formal gardens – grown in rows as an allée or paired to frame a garden feature.

Cupressus sempervirens is a quick-growing evergreen, the dense, dark green, scaly waxy foliage held on ascendent branches. Rounded, roughly golf ball-sized seed cones form on branches turning from green to brown as they age, and if produced in quantities, pulling branches slightly down from the vertical. Plants are native to the eastern part of the Mediterranean and western Asia; trees in wild populations are more variable than seen in gardens, with individuals of a broader, conical shape rather than pencil-like form and easily reaching 30m, sometimes more, while remaining perhaps 3m across.

This durable tree has been cultivated since antiquity. It may have been planted in early Assyrian gardens; stone reliefs found at Nineveh in Northern Iraq from around 700 BCE show superficially similar, slender, conifer-like trees (and the city is believed by some scholars to be the site of the legendary Hanging Gardens of Babylon). In Abarkuh in neighbouring Iran, there survives an example alleged to be 4,000 years old; certainly, these trees do prove to be impressively long-lived once established, tolerating summer heat and drought with ease.

In cooler, temperate climates, the tree increasingly proves successful as our climate warms and long, freezing winters are less commonplace. It stands some frost well, undamaged by temperatures of around -10°C and surviving -15°C. Choose as sunny a place as you can find, somewhere not overshadowed by taller trees and with really well-drained soil – this can be fairly poor and stony. It does, however, grow best in

Opposite Italian cypress (*Cupressus sempervirens*) by Mary Anne Stebbing, 1893. Kew Collection.

Opposite Italian cypress (*Cupressus sempervirens*) from *Traité des arbres et arbustes qui se cultivent en France en pleine terre* by Duhamel du Monceau, 1806.

some shelter – winds can damage its slender form while heavy snowfall can pull down branches, spoiling the shape; in cold districts, it does best tucked in beside a warm wall. For good results, plant young trees, perhaps no more than around 60cm high – these will grow quickly and strongly; taller specimens will need staking until their roots establish, which utterly spoils their appearance and makes the extra expense pointless. Newly planted trees need watering well for the first year.

Plants look superb in gravel gardens, soaring from between the rounded forms of laurels, olive trees or smaller *Cistus* and *Phlomis*, perhaps softened by clumps of *Stipa gigantea* or *Cortaderia* and adding a sense of rhythm when repeated through flowing naturalistic planting – they are especially impressive seen at different heights, perhaps arranged on a bank, hillside or terraced slope. In formal gardens they also make a great contribution, perhaps best planted as pairs or alongside clipped topiary shapes. In some cultures, it is even seen as unlucky to use them in odd numbers. *Cupressus sempervirens* remain impressive features year-round – especially dusted with frost or draped with dewy cobwebs. Snow is a danger, however, and should be knocked off if it settles. Any branches that are pulled down out of position, spoiling the shape, can usually be tucked back in if attended to promptly.

OTHER SLENDER CONIFERS TO TRY

- ***Cupressus sempervirens* 'Green Pencil'** is a reliably narrow selection with rather bright green foliage; **'Swayne's Gold'** by contrast has bright gold foliage; it is slower-growing than green selections but makes a great contrast with them and may develop orange tints in winter. It may also be slightly less hardy. **'Totem'** has dark green foliage and is said to spread to no more than 1m across.
- ***Juniperus scopulorum* 'Skyrocket'** makes a good alternative to ***Cupressus sempervirens*** in particularly cold locations because it is similarly slender with greyish-green foliage and will stand drought when established as well as temperatures of -20°C given a sunny, well-drained site. Similar **'Blue Arrow'** has more glaucous foliage.

T. 3. N. 1.
CUPRESSUS sempervirens.
CYPRÈS ordinaire.

Ivy-leaved cyclamen

Cyclamen hederifolium

Versatile, low-growing perennial for sun or shade adapted to seasonal drought, bearing superb displays of autumn flowers and, in winter, handsome carpets of silver embellished foliage.

Opposite Ivy-leaved cyclamen (*Cyclamen hederifolium*) from *Curtis's Botanical Magazine*, 1807.

This tuberous perennial is known as one of the most adaptable and reliable of garden plants, thriving in a range of conditions, from arid, open and sun-baked ground to moist, sheltered shade, and everything in between, as long as soil is well drained. As a result, *Cyclamen hederifolium* is a great candidate for dry sites that experience summer drought but also happen to be shaded, perhaps below shallow-rooted trees and shrubs or in narrow borders above dry footings of shaded walls. It is another Mediterranean native, found naturally in the wild in woods, scrub and on rocky ground, as far west as the south of France and east to parts of Turkey; it makes a particularly splendid sight when it flowers in autumn in the olive groves of southern Europe. This cyclamen has naturalized in many areas of temperate northern Europe, where its range is increasing as a garden escape.

This species' physiology and life cycle are perfectly adapted to surviving dry conditions, the foliage and flowers sprouting in late summer and early autumn just as rains and cooler weather arrive. Growth emerges from a broad, fleshy, water-storing tuber, which may reach the size of a dinner plate in time. The flowers are generally produced first and are around 2cm long, held on stalks to around 10cm, occasionally more. The handsome, vaguely ivy-like foliage (hence the Latin name) arises a little later, leaves reaching around 10cm long in some individuals. They are green and variably heart-shaped and attractively patterned in silver – the best of these, as well as occasional pure silver-leaved selections, are highly prized by gardeners. Established drifts of this cyclamen with its overlapping foliage can form most attractive low winter ground cover. The charming white, pink or reddish flowers with their five reflexed petals are produced in great quantities from mature plants. These are self-fertile, allowing plants to set seed freely. This develops in rounded capsules attached to the tuber by the old flower stalk, which becomes curiously coiled and spring-like, drawing the ripening pods close to the tuber. The capsules eventually burst open when fully ripe to reveal black, rather sticky coated seeds. This sugary coating is irresistible to ants, who take them back to their nests, effectively dispersing the

Opposite Ivy-leaved cyclamen (*Cyclamen hederifolium*) from *Edwards's Botanical Register*, 1883.

seed. As a result, seedlings pop up all over the garden, often to charming effect. By the time seeds have been dispersed, the foliage will have withered away and the plant enters summer dormancy to survive the hottest, driest part of the year. In winter it is perfectly hardy to around -15°C.

In gardens this species proves useful in a range of situations. It thrives in sunny or shaded rock and scree gardens, enjoying free-draining conditions and makes a good choice for underplanting in gravel gardens. The ideal place is ground that will not see disturbance – anywhere that is dug over should be avoided, as tubers are easily speared by tines of a garden fork. In woodland gardens and below shrubs it makes considerable impact in autumn beside goblets of *Colchicum* and delicate autumn-flowering *Crocus* and *Galanthus*, adding spring-like freshness as elsewhere leaves turn crimson and gold. It's also perfect in tree circles in a lawn – that area directly below a specimen tree where grass struggles to grow – and is also a candidate for tucking in at the dry base of a hedge – plants even thriving and flowering at the foot of evergreen *Taxus* (yew). Gardens where it thrives are often also home to similar, winter- and spring-flowering *Cyclamen coum*, a daintier, smaller plant – take care that *C. hederifolium* does not outcompete if grown nearby.

OTHER CYCLAMEN TO TRY

- ***Cyclamen coum*** is almost as easy and arguably better in flower, blooming in winter and spring and thriving in similar places as ***C. hederifolium.*** Leaves are smaller and daintier and it is less vigorous. Height 5–7cm. Hardy to around -15°C.
- ***C. cilicium*** is a charming plant, flowering in autumn with soft pink or white, scented flowers and silver marked leaves. It likes a sheltered, sunny place or makes a small pot in an alpine house. Height 7cm. Hardy to around -5°C.
- ***C. graecum*** (Greek cyclamen) bears pink or white, often scented flowers in autumn above beautifully marked, heart-shaped foliage. Forms a taproot and needs a sunny, sheltered place outdoors or a deep pot in an alpine house. Height 7cm. Hardy to around -5°C.

Trailing ice plant

Delosperma cooperi

This low, ground-covering succulent plant from South Africa stands extreme summer heat and drought, producing masses of shimmering mauve flowers, and if kept dry in winter will also tolerate considerable frost.

From South Africa comes this low-growing, potentially tumbling succulent, which is a rather glamorous plant when in flower, yet one that proves easy to keep. It is surprisingly hardy given good conditions, and is increasingly used in cool, temperate regions as summers become warmer and more drought-prone, serving as a good replacement for more traditional *Aubrieta* or *Aurinia saxatilis* (golden alyssum) where sites may get too dry and hot and a trailing plant is needed.

In the wild it grows in places with minimal competition, thriving in poor, rocky soil, sometimes at considerable altitude clinging to cliffs and scree. Plants reach about 6–8cm high, but the slender trailing freely branching stems can sprawl much further, forming a low quite dense mat perhaps 1m or more across. The narrow, cylindrical leaves are around 3cm long and covered in tiny silver hairs, which gives the whole plant a glistening look. The highlight, of course, are the flowers – these are usually vivid pink-magenta and daisy-like with dozens of slender shimmering petal-like ray florets, and around 4–5cm across. They are produced in considerable profusion all summer, covering the plant at times, a habit that has given rise to the plant's alternative common name of pink carpet.

For dry gardens with a Mediterranean climate that experience only light frost, this plant could not be simpler to keep. In urban areas it will make useful low-care ground cover, while it especially thrives in coastal districts, even on sheer banks, walls and rock cliffs with full maritime exposure. In more temperate regions this plant's tolerance of both high summer heat and considerable winter cold has resulted in its recent popularity. Plants are hardy to an impressive -15°C, possibly rather lower, but with the proviso that they are kept quite dry – cold and wet conditions will quicky finish the plant off, so make sure *Delosperma* is planted somewhere really well drained, ideally in a raised position to ensure that water can run through the roots freely. Mulch the soil surface with grit or gravel to draw any moisture away from plant stems. It may even be worth covering plants with a sheet of Perspex in winter to keep off the worst of the wet.

Opposite Trailing ice plant (*Delosperma cooperi*) from *Curtis's Botanical Magazine*, 1877.

Opposite *Delosperma sutherlandii* from *Curtis's Botanical Magazine*, 1877.

A dry, sunny, retaining wall is a perfect spot to try it; here the stems will cascade to superb effect. Otherwise try spilling it down a dry bank, or from a pocket in a rock garden, or let it carpet the very front of a hot border. In a narrow planting space at the dry base of a wall the procumbent stems can spread along and enjoy any reflected heat – these stems will grow out over paving too, helping to soften hard landscaping in sites such as a sunny terrace. Otherwise grow it in containers of sharply drained compost with plenty of added grit; planted Belfast sinks are perfect, shallow terracotta pans displayed on outdoor tabletops or atop low terrace walls also work well, as do wall planters where flowering stems can be allowed to drip over the edges. Containers are, of course, easily moved under cover for winter if need be. The other key consideration is to find *Delosperma* a place with little competition – overhanging vegetation casts shade, and this plant does not like much.

You can plant it with, or at least around, other succulents – hardier *Agave* and *Aloe* – or with fleshy-leaved alpines such as *Lewisia*, *Sempervivum*, even some smaller *Sedum* and *Saxifraga*. It can also be easily combined with other similar ice plants, such as some *Carpobrotus* (sour fig) or yellow-flowered *Delosperma nubigenum*, which form a low mat around other plants to great effect. It also makes a good fringe in front of perennials such as *Agapanthus*, *Kniphofia*, bearded *Iris* or even bulbous, autumn-flowering *Nerine*.

OTHER *DELOSPERMA* TO TRY

- ***Delosperma congestum*** has large yellow flowers each with a paler eye, held above neat, rather dense green, carpeting foliage. Hardy to -10°C, perhaps lower if dry in winter.
- ***D. cooperi*** Jewel of Desert Series comprises a range of selections with different coloured flowers including red, yellow and orange-pink. Hardy to -10°C if dry in winter.
- ***D. nubigenum*** is another yellow-flowered, carpeting species, with masses of small blooms all summer on a vigorous wide-spreading plant. Hardy to -10°C, perhaps lower if dry in winter.
- ***D. sutherlandii*** is compact growing, with large violet-pink flowers. Hardy to -10°C if kept dry.

1
1.
1

Tree poppy

Dendromecon rigida

Bearing a succession of golden poppies throughout summer, this impressive shrub makes a great choice for training on to a sunny wall, where it will stand winter frost and summer drought well.

Opposite Tree poppy (*Dendromecon rigida*) from *La Belgique horticole*, 1860.

From the region of northern California and into Mexico comes this distinctive and highly appealing woody member of the poppy family, which grows in chaparral and on sharply drained dry, rocky, sometimes wooded slopes in the foothills of mountain ranges, including the western Sierra Nevada. Tree poppy is a rather twiggy rounded evergreen shrub, reaching in the wild around 3m tall with pointed, narrow, rather rigid, lance-shaped, blue-green toothed leaves, to around 10cm long. The stems bear from spring through early summer and later on into autumn a succession of shining, golden-yellow, usually four-petalled flowers to around 5cm across – fragrant and attracting pollinators, they are held singly on slender stems and stand out beautifully from the plant's glaucous foliage, often produced in considerable numbers for an impressive display. Individually flowers look remarkably similar to those of annual *Eschscholzia* (California poppy) which can be found growing in similar habitat. This tree poppy was introduced to UK cultivation in 1854 but has never been common; it is closely related to another rather more flamboyant shrubby poppy, *Romneya coulteri*, also from California. It shares with that species a liking for areas that have recently suffered wildfires. There is also another species in the genus, *D. harfordii*, which is native to the Channel Islands of California and is a taller-growing plant with broader, more oval foliage and larger flowers; it is still rare in cultivation but may fit well at the back of planting in a relaxed naturally-styled gravel garden.

In gardens, tree poppy is a highly prized plant. Tricky to propagate, young plants are never plentiful and always in demand. It is, however, easily grown once established, thriving in full sun and well-drained, even quite poor, sandy soil, and enjoying the conditions in many milder gardens in Mediterranean climates where summer drought is of little concern to it. Tree poppy is not fully hardy, but survives at least -5°C, probably less. In frost-prone temperate regions, plants may become semi-evergreen and are most often seen wall-trained, the branches tied flat on to wires to fine effect. This actually shows the plant off to best effect: free-standing plants can be rather shapeless and hard to manage.

DENDROMECON RIGIDUM Benth.

Opposite Tree poppy (*Dendromecon rigida*) from *Flore des serres et des jardin de l'Europe*, 1816.

Growing in this way also allows the tree poppy to catch as much sun and heat as possible, while the rain shadow effect of the wall helps to keep it on the dry side; plants grown on a wall will also usually stand all but the coldest winters. Tree poppy is not a plant that likes too much direct competition. It will, however, grow well beside other sun-loving wall shrubs such as *Ceanothus* or *Solanum crispum*, which both have blue flowers that contrast beautifully with the golden blooms of the poppy early in the season. Later on, scented yellow-flowered *Cestrum parqui* or *Lespedeza thunbergii* with its pendulous purple flower racemes prove an appropriate neighbour. You could try scrambling a blue-flowered clematis such as outstanding 'Perle d'Azur' or 'Prince Charles' through its branches to contrast with those golden blooms, just as long as the poppy does not become overpowered. At its feet, sun-loving *Nerine* or bearded *Iris* make a good choice.

OTHER WALL SHRUBS TO TRY

- ***Cestrum parqui*** (willow-leaved jessamine) is a deciduous shrub that needs a hot, sunny wall and stands dry conditions. It bears masses of starry greenish-yellow flowers in summer that are sweetly scented at night. Height 2.5m. Hardy to around -5°C.
- ***Fremontodendron californicum*** is another sun-loving, drought-tolerant evergreen wall shrub native to California. It too bears masses of showy yellow flowers, around 6cm across through summer and into autumn. Its lobed foliage bears irritant hairs, so take precautions when handling. Needs a tall wall. Height 4m. Hardy to around -10°C.
- ***Lespedeza thunbergii*** makes a spectacular sight in late summer when its arching stems cascade with pendent purple flower heads. Likes a hot wall and a well-drained site. Height 2m, hardy to around -15°C.
- ***Solanum crispum*** (potato vine) bears bunches of yellow-eyed purple flowers all summer on a warm sunny wall, surviving drought when established. Height 4m. Hardy to around -10°C.

African daisy

Dimorphotheca jucunda (syn. *Osteospermum jucandum*)

Marvellous daisy-flowered evergreen perennial ideal for ground cover in sun, or for spilling over a hot, dry bank or low wall, bearing masses of blooms all summer.

Opposite African daisy (*Dimorphotheca jucunda*) illustrated by Malcolm English, 2024.

This enchanting and useful daisy-flowered, ground-covering perennial is native to parts of southern Africa, including the Eastern Cape and Natal. Here it is commonly found carpeting free-draining slopes of hills and mountainsides, as well as upland grassland areas. It is one of the many sun-loving daisies from that part of the world which, often grown as bedding and seasonal plants, help bring much colour and joy to frost-prone, temperate gardens during summer. This *Dimorphotheca*, however, often proves reliably hardy, performing with gusto from year to year over a remarkably long season if conditions are right. Plants form a dense mat of growth to around 15cm high, from a mass of procumbent, branching stems that lie on soil surface and often root in as they sprawl, a habit that helps the plant escape unsuitable conditions should they develop. The long, rather slender leaves are pale grey-green, around 8cm long and rather rough to touch, and slightly aromatic.

The flowers of *Dimorphotheca jucunda* are delightful; around 6cm across when fully open and produced generously, held on long, slender flower stems that are rather sticky to touch. They react quickly to sunlight, opening promptly on sunny days but often remaining shut when overcast. Most commonly the daisies have a pinkish-mauve, petal-like ray floret with a darker central eye, but the flowers can vary from near-white to rich purple. Plants prove fairly tough customers, surviving summer drought well and yet tolerating cold spells, which it often encounters in its natural range; plants are hardy to around -5°C, probably lower for short spells. In some countries with mild winters, it has naturalized as a garden escape, such as in parts of Ireland.

In sunny gardens *Dimorphotheca jucunda* is a quick-growing plant – it does especially well in open sites that are not too cold or wet in winter and in gardens near the sea because it thrives on maritime exposure. It does best in full sun, and needs fairly sharp drainage; a dry slope makes the perfect place, but a rock garden or the edge of a raised bed is also highly suitable. The stems will also travel admirably across gravelled areas or spill beautifully over the edge of a path or a paved terrace. It is often

Opposite
Dimorphotheca ecklonis from *Curtis's Botanical Magazine*, 1897.

the case in gardens that over a few years shade develops where once was sun. This plant tolerates these changes well, the stems growing towards the sun and rooting in. It actually stands some shade, especially in gardens with a hot Mediterranean climate, although flowering may be reduced, which may alert the gardener to act and replant somewhere better suited. If winter has been mild, flowering begins in mid to late spring in a sudden flush of dazzling bloom; after colder conditions growth will have been burned back and flowering is delayed until new shoots develop.

To keep plants looking good and to promote more flowering, deadhead from time to time, removing the long flowering stem where it joins the main plant. In time, individuals may need trimming back, dead branches removing or starting again from cuttings or rooted sections of stem – an easy task best tackled in spring.

Dimorphotheca jucunda makes a great carpet in front of *Yucca*, *Agave* or *Cistus*, while bulbs such as larger *Gladiolus* can look good emerging through it. Let it tumble down banks beside *Erigeron karvinskianus* or succulent *Carpobrotus*. There are several selections worth looking for. With almost white flowers 'Lady Leitrim' is one of the most reliable, while for intense colour choose *D.* 'Stardust' probably a hybrid, with purple-pink flowers from early in the season.

OTHER DAISIES TO TRY

- ***Arctotis* × *hybrida*** are admirably suited to dry hot conditions, forming attractive low mounds of silvery foliage and attractive daisies in often fiery colours. They are great on patios in containers but must be overwintered under glass in temperate regions. Survives temperatures down to 0°C.
- ***Dimorphotheca ecklonis*** is a shrubbier, more upright species than ***D. jucunda*** and a parent of many garden hybrids, with flowers in a wide range of colours and sometimes variegated foliage used as summer bedding in hot, dry places. Hardy to 0°C.
- ***Gazania rigens*** bears dazzling daisy flowers in shades of yellow, orange and red, often with contrasting dark spots in the centre of the flower. It blooms all summer in a sunny place above tufts of silvery-green foliage and has been widely used to develop hybrids for bedding purposes. Hardy to 0°C.

2
4
3
1.
5

Mexican snowball

Echeveria elegans

Forming wonderful rosettes of grey-blue succulent leaves, this compact-growing plant is easy to keep and quick to bulk up. It thrives in frost-free environments and makes an easy summer succulent for a pot in a sunny corner in colder gardens.

Opposite Mexican snowball (*Echeveria elegans*) illustrated by Malcolm English, 2024.

Few succulent plants are easier to accommodate than this delightful little *Echeveria*, native to arid parts of central Mexico, often in mountainous areas where it grows in crevices and on rocky outcrops. They closely resemble hardy *Sempervivum* (houseleek), plants forming tight rosettes around 10–15cm across of small, succulent, overlapping spatula-shaped leaves. In *Echeveria elegans* these are blue-white but occasionally, in really strong sun, pink-tinged, each with a small blunt point at the otherwise rounded leaf tip. Plants increase freely by producing small offset rosettes which crowd tightly round the mother plant and give this succulent its other common name: hen and chicks. Plants in time form clusters of rosettes that encrust rocks or spread out over the soil, forming dense clumps to around 30cm or more across. In late winter and spring, from the centre of the rosettes arise arching flesh-pink flower spikes reaching around 25cm tall and hung with pink, yellow-tipped, bell-shaped flowers. While *Echeveria elegans* occasionally encounters light frosts in its native habitat, plants are generally regarded as tender, standing just occasional dips to perhaps -3°C if kept perfectly dry.

In subtropical or Mediterranean gardens, these compact succulents are among the most reliable and easily grown of all drought-tolerant plants, thriving atop or even growing from the face of a retaining wall, flourishing below larger succulents and cacti as ground cover or spilling from terracotta pans and planters. They need next to no care, virtually no soil and stand heat and drought brilliantly, while the silvery blue of their foliage complements and contrasts well with a host of other plants. They are useful too in cooler, temperate climates. Being quick to propagate via rooting the numerous offsets produced, *Echeveria elegans* is often used as a sort of succulent bedding plant – planted out in rock gardens or even used in intricate carpet bedding schemes for summer. They are most easily kept as container plants placed outdoors on a sunny terrace or patio for the frost-free months, then overwintered in a greenhouse or on a sunny windowsill inside. Grow them in small single pots or wider pans. Alternatively, combine them in a large gravel-topped dish (make sure it has good drainage holes) with a collection

Opposite *Echeveria agavoides* from *Flore des serres et des jardin de l'Europe*, 1873.

of other small near-hardy succulents such as *Aristaloe aristata, Delosperma cooperi, Haworthia, Gasteria, Crassula* or *Aeonium*; this makes a superb, low-maintenance tabletop display for a sunny outdoor dining area in summer, and if you have a conservatory it can be enjoyed through winter too.

During the colder months under cover, *Echeveria* need as much light as they can get to prevent them becoming etiolated and more susceptible to rotting. Give them next to no water at all, unless the foliage starts to shrivel, waiting until spring to give pots a soak before moving them back outdoors. As they bulk up so freely, it is worth experimenting with sacrifical spare offsets, planting in sheltered spots where they might manage to overwinter. Plants can survive surprisingly well for years if clumps stay dry, although eventually a hard winter – particularly one with snow – may see them off. In an especially sheltered site outdoors, perhaps in an urban area, they are worth trying as part of a planted green roof because they need only the thinnest layer of substrate to flourish.

OTHER *ECHEVERIA* TO TRY

- ***Echeveria agavoides*** is a rather less hardy, slightly larger species with more pointed green leaves, making a plant that looks rather like a miniature ***Agave***. There are many named selections, such as 'Ebony', which has red-tipped leaves, creating a startling appearance. Best above 5°C.
- ***E. affinis*** also has pointed foliage, but this time the leaves are dark green-brown, almost black. Bears red flowers. Best above 5°C.
- ***E. lilacina*** (ghost echeveria) has pale silver-grey foliage that forms an impressive rosette, 17cm wide. It is slower-growing but worth the wait. Best above 5°C.

ECHEVERIA ACAVOIDES *Lem.*

California fuchsia

Epilobium canum
(syn. *Zauschneria californica*)

This late-flowering, low-growing hardy perennial produces masses of usually red, tubular flowers from late summer into autumn. It loves the sun and stands dry conditions well.

Opposite California fuchsia (*Epilobium canum*) from *Curtis's Botanical Magazine*, 1850.

At the end of a long hot summer, this low-growing, shrubby perennial provides a great show with a sudden flash of rather exotic, fiery flowers, which often appear just as the first autumn leaves are starting to fall. California fuchsia is a native of California and Mexico, where it grows on slopes and in chaparral, often in coastal areas. The rather tubular, usually scarlet- and orange-tinted flowers are 3cm long and clustered at shoot tips; each is flared at the mouth with four petals and look like those of some *Fuchsia*. Rather marvellously, these blooms attract hummingbird pollinators in the wild, hence the plant's alternative common name: hummingbird flower. *Epilobium* is a member of the willowherb family, which includes not just *Fuchsia* but also the genus *Oenothera* (evening primrose), many of which are great plants for dry, sunny gardens. *Epilobium canum* is a rather variable species with a couple of subspecies, but all form mounds of growth with freely running roots that travel widely, colonizing rocky places, basking in sun and heat and free-draining soil, reaching around 60cm high and more across. The leaves are generally slender, lanceolate and slightly hairy, often soft green in hue, although foliage can be distinctly silvery in some plants, making an unusual contrast with the flowers, which are carried in great profusion and appear in the wild at a time when little else is in bloom, the display lasting a month or more. Plants are generally hardy to around -10°C given good drainage, although they will behave as herbaceous plants in colder winters.

In cultivation, *Epilobium canum* is often still sold under the old name of *Zauschneria*. It is a reliable, easily grown plant for gardens with a Mediterranean climate; it resists drought well once established, and flowers wonderfully after a long hot summer, needing little care other than a trim back at the end of the season. Plants spread about by runner-like rhizomes, which means new shoots can pop up some distance from the original plant; helpfully, the rather brittle shoots tend to appear rather late in spring or early summer, rapidly filling any remaining gaps in planting. Suckers are easily removed – if you can extract them with a bit of root and pot them up promptly, they quickly make new plants.

Opposite California fuchsia (*Epilobium canum*) from *Flore des serres et des jardins de l'Europe* by Louis van Houtte, 1848.

In cooler, temperate regions, the plant's relatively low-growing habit and liking of free-draining soil makes it suitable for sunny rock gardens, but this plant will also thrive in a wide range of other situations. It can look great filling a narrow border at the foot of a sunny wall, perhaps following on from spring and early summer bulbs. It also makes a vibrant and unusual front-line planting in herbaceous borders, where it will obligingly soften edges and spill gracefully over a path and even run into gaps between slabs. It is an excellent plant for a gravel garden, where its late colour is appreciated, and can also be planted in raised beds or at the top of a retaining wall, or else allowed to grow down a slope – all places where the drainage will be sharp.

The late flowers often attract pollinating bees. Frost will quickly end the flowering period, but this may be as late as November, which means *Epilobium canum* can be used in partnership with some other autumn performers. Allow it to grow beside *Cyclamen hederifolium* or let the goblets of *Colchicum* or pink spidery blooms of *Nerine* to emerge nearby; use it in front of succulent clumps of *Hylotelephium*, below daisy blooms of various asters or more luxuriant Japanese anemones. It also goes well in front of grasses such as *Cenchrus longisetus* and *Nassella tenuissima*, and mingles delightfully with obliging *Erigeron karvinskianus.* Once flowering is finished and the top growth fades, simply cut it down to ground level and your work is done.

OTHER SELECTIONS TO TRY

- ***Epilobium canum* 'Albiflorum'** is an unusual, clean-white flowered, green-leaved plant, great for using as a contrast with usual red-orange selections. Height 30–40cm.
- ***E. canum* 'Dublin'** bears intense scarlet-red flowers held over small green foliage. Height 40cm.
- ***E. canum* 'Olbrich Silver'** is an outstanding American selection with really silver foliage which sets off the scarlet trumpets perfectly. Height 40cm.
- ***E. canum* 'Solidarity Pink'** has soft pink flowers, which show up well against the plant's pale green leaves. Height 30cm.

ZAUSCHNERIA CALIFORNICA *Presl*

Mexican fleabane

Erigeron karvinskianus

Bearing countless daisy flowers from mid-spring to early winter, this tough little sun-loving perennial is one for gardeners who like a relaxed look, with plants self-seeding freely into paving cracks and old walls.

Opposite Mexican fleabane (*Erigeron karvinskianus*) from *Curtis's Botanical Magazine*, 2012.

If a dynamic, almost non-stop display of cheerful daisy flowers sounds like the characteristic of a great garden plant, then this marvellous, little hardy perennial will appeal. Unsurprisingly perhaps, this native of Mexico, as well as other parts of Central America, has a liking for sun, warmth and dry, well-drained conditions. *Erigeron karvinskianus* is also one of the easiest drought-tolerant plants to grow, perhaps too easy in some cases, spreading exceptionally freely by seed. It is widely naturalized in some regions, be they tropical or temperate, even becoming invasive if conditions are to its liking – in Hawaii and New Zealand for example. In many plots great or small, however, it is welcomed, even in some of the most celebrated gardens of all, sprouting from cracks in paths and terraces, and forming a wonderful froth of daisies enlivening and softening steps or draping old walls all summer.

The common name of fleabane comes from the once-held belief that *Erigeron* repels fleas, but in gardens this proves a great plant for attracting pollinating insects, especially hoverflies. While this *Erigeron* is best in sun, plants often self-seed and then thrive in shade too, just as long as drainage is good. Individual plants are not long-lived, surviving three to five years perhaps, but there is always a succession of younger seedlings ready to take over. Summer heat and spells of drought seldom bother this plant, and winter cold has little effect either, plants surviving at -20°C.

Individually the flower heads look much like a refined lawn daisy, with a yellow centre but slightly finer, petal-like ray florets that are usually white with a pinkish tint that intensifies as flowers age. About 1.5cm wide, these are held atop long, wiry flower stems that lift the blooms above the plant. In exposed sites, such as by the coast, these flowers are almost always in motion, which is why the plant is sometimes known as 'dancing daisy'. Growth on established plants can be rather shrubby with masses of floppy shoots bearing small, slightly hairy, lance-shaped bright green leaves. In all it makes a sprawling mound perhaps 20cm high and 30cm across on larger plants, the lower stems rather woody at the base. Often the flowers can be produced more or less year-round: in most temperate gardens,

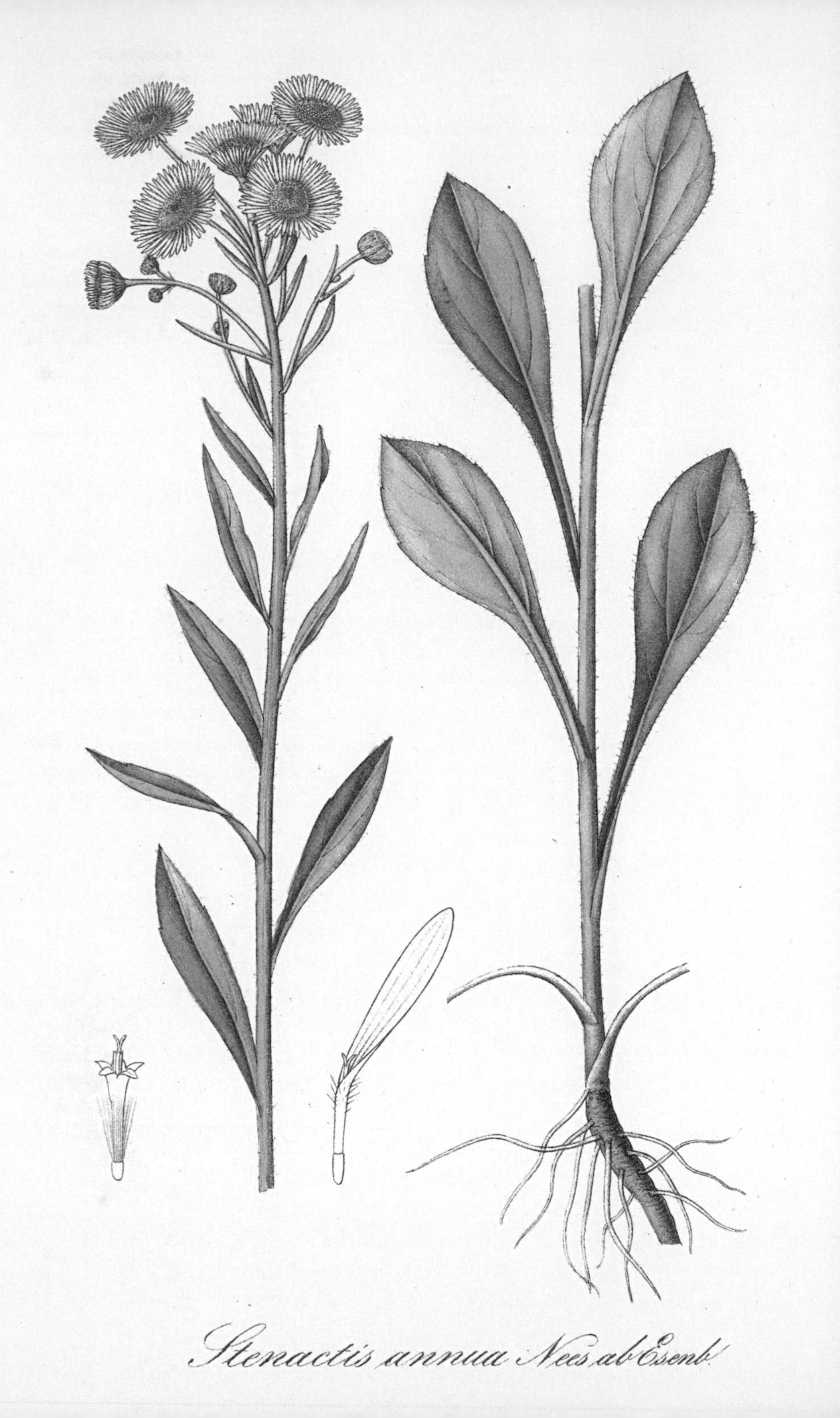

Stenactis annua Nees ab Esenb.

Opposite *Erigeron annuus* from *Flora Regni Borussici* by Albert Dietrich, 1843.

the display usually starts by mid-spring, the plants providing a seemingly endless succession of flowers through summer regardless of the conditions, remaining in good shape into autumn, an invaluable characteristic. The first hard frosts of winter will put paid to them, if by then plants have not already been trimmed back, for by now they may be bedraggled and choked with leaves and other debris and are best sheared off at just above ground – or paving – level; new shoots appear in early spring.

The best and longest-lived plants are always those that have self-sown and which spring from potted plants and cracks in hard landscaping – perhaps from the crumbling mortar of a wall or between paving slabs, spilling out over a terrace or at the side of a path. This is really the best place for them – *Erigeron karvinskianus* will do well in rock gardens too but can quickly swamp slower-growing neighbours and become an annoyance. Getting plants established in the first place is sometimes tricky; bought plants tend to be short-lived once planted out. If you do so, choose a really well-drained place or grow in a container by the area you want to colonize and allow it to set seed – do not be too fastidious with weeding. Alternatively, in spring, weed out your paving cracks, pop in a little seed compost and sprinkle some seeds on top. As soon as seedlings appear, you know you've got it for life.

OTHER *ERIGERON* TO TRY

- ***Erigeron annuus*** is similarly free-flowering with masses of small white daisies all summer, held in clusters on 1m upright stems. Likes full sun. Hardy to at least -10°C.
- ***E. aureus* 'Canary Bird'** makes a small, neat perennial for a sunny rock garden or alpine trough. In summer yellow daisies held on stems to 6cm tall appear. Hardy to -10°C.
- ***E. glaucus*** is common in maritime gardens, forming mats of blue-green foliage topped off by pink or mauve flowers 6cm wide. Hardy to around -15°C.

California poppy

Eschscholzia californica

This quick-growing annual or short-lived perennial will dazzle in summer with long-lasting shimmering displays of poppies that prove highly resistant to heat and drought.

Some plant species survive stressful conditions by being quick-growing and short-lived, maturing rapidly and flowering, then enduring difficult growing conditions as seed, a technique often displayed by drought-tolerant annuals and short-lived perennials. Native to an extensive sweep of the western United States (particularly California, where it is the State flower) as well as Mexico, lovely *Eschscholzia californica* comes from grassland and other open habitats and has naturalized in countries with similar climates, such as Chile and Australia. California poppy is usually a short-lived perennial in the wild, although in dry, hotter (as well as colder) parts of its range it is an annual.

At flowering time, this poppy can form incredible sheets of blooms, the flowers extending as far as the eye can see, usually in orange but also yellow and occasionally other colours. The cupped flowers are composed of usually four satin-like, rounded petals arranged around a central crown of golden stamens, in all around 7cm wide. These are held singly on long stems above the plant. Flowers remain closed in overcast conditions and shut at night, but on sunny days they open almost flat in succession for weeks. Small seeds develop inside a finger-like capsule which bursts open when ripe, scattering contents. The plant is usually 20–30cm high, occasionally more, the attractive foliage blue-green and finely divided.

Plants love sun and well-drained, poor, sandy soil and thrive in exposed places and maritime districts; they are among the simplest of plants to grow. *Eschscholzia* flower best in a hot summer and seldom need irrigation, standing drought with ease. In temperate climes, California poppies are usually treated as annuals, although in mild years plants survive over winter. Since plants do not like root disturbance, seed is direct sown after the risk of frost has passed, in late spring or early summer. They bloom generously from midsummer, well into autumn, especially if you deadhead – if not, they self-seed freely, numerous young plants arising in spring, which are easily edited if unwanted. Pollinating insects such as bees and hoverflies love them: the flowers are particularly pollen-rich, so often recommended for use in wildlife gardens. Grow in gravel or rock gardens where

Opposite California poppy (*Eschscholzia californica*) from *Edwards's Botanical Register*, 1835.

Opposite California poppy (*Eschscholzia californica*) from *Paxton's Magazine of Botany, and Register of Flowering Plants* by Joseph Paxton, 1837.

they spill out over gravel and rocks and soften bold architectural plants such as *Agave* and *Yucca*. You can sow them at the front of borders for a colourful low edge; use them to fill gaps where spring bulbs such as daffodils and tulips have faded; scatter with seed of other annual flowers such as cornflowers in meadow-style plantings; or, at the other extreme, simply drop seeds in pots for drought-tolerant container displays. They even cover ground below roses if in full sun.

There are many selections, some with single or double flowers in shades other than the usual bright orange and yellow, including near-white ('Ivory Castle'), peach ('Peach Sorbet'), pink ('Apple Blossom Pink'), purple ('Purple Gleam') and even vivid red ('Red Chief').

OTHER ANNUALS TO TRY

- ***Calendula officinalis*** (pot marigold) is easy to grow in a sunny place with masses of pollinator-friendly, orange or yellow daisy-like flowers from early to midsummer. Direct sow outdoors after frosts, or sow in modules for earlier blooms, keeping moist until established. Self-seeds. 50cm.
- ***Limnanthes douglasii*** (poached egg plant) forms a mat of growth that is smothered with cupped white-edged, yellow flowers in summer. Direct sow in spring in a sunny place where you'd like it to flower. Self-seeds. 15cm.
- ***Nigella damascena*** (love-in-a-mist) is an upright plant with blue, pink or white flowers in early summer and masses of soft, frothy foliage. Direct sow in sun or sow in modules and plant out, keeping moist until established. Self-seeds. 50cm.
- ***Papaver lecoqii* 'Albiflorum'** (Beth's poppy) is an upright poppy with pale pink flowers in summer, ideal for a hot, sunny site. Sow in modules and plant out, keeping moist until established Self-seeds. 40cm.
- ***Portulaca grandiflora*** (moss-rose purslane) is a low succulent plant with rose-like, often double flowers in many shades. For full sun and thrives in the driest spots. Sow in modules and plant out after frost. 8cm.
- ***Tithonia rotundifolia*** (red sunflower) is grown as an annual – it bears orange-red flowers that look like dahlias, flowering from summer into autumn and attracting butterflies. Sow in modules and plant out after frosts, keeping moist until established. 1.5m.

Eschscholtzia crocea.

Narrow-leaved black peppermint

Eucalyptus nicholii

This handsome and hardy gum makes a fairly compact, rounded tree with weeping branches bearing aromatic blue-green foliage and then in autumn masses of fluffy white flowers.

Opposite
Narrow-leaved black peppermint (*Eucalyptus nicholii*) illustrated by Malcolm English, 2024.

Is there a genus of trees more redolent of the Australian outback than *Eucalyptus*? Probably not. With their distinctive, greenish-blue, aromatic evergreen foliage, gum trees are synonymous with their native land, providing a distinct character that goes hand in hand with sunbaked orange-red soil, billabongs and koalas. In truth *Eucalyptus* is a highly varied genus of around 700 species, including the mightiest flowering plant of all, sky-scraping *E. regnans* (mountain ash) at 100m tall (possibly more before loggers chopped down the tallest trees, most of which was done in the nineteenth century) all the way down to mountain-dwelling, multi-stemmed Tasmanian *E. vernicosa* (varnished gum), which reaches just 1m.

Members of the myrtle family, *Eucalyptus* are widely cultivated commercially in plantations around the world, due to their fast-growing habits, and in many cases a tolerance of hot, dry conditions. They are also enjoyed in gardens for their often statuesque appearance. One curious habit is that many have distinct phases of growth – young plants bearing foliage markedly different in shape and sometimes colour from adults. The leaves often hang down from branches so that shade cast by trees is surprisingly light. Many are adapted to survive wildfires, resprouting from old growth or with seeds that germinate freely in burned ground. The flowers are distinctive too – they have no petals but instead are a mass of fluffy stamens – usually white, but also in pink, red or yellow.

Lovely *Eucalyptus nicholii* is found in the wild in only a few small localities in New South Wales, but it is regarded by many as one of the most handsome of all for gardens, reaching 15–18m, pretty compact for a gum tree. It is one of several which have foliage that smells distinctly of peppermint if crushed, and in the wild grows in poor, rather shallow soil, in grass or woodland. While not common (it is regarded as 'vulnerable' by the Australian government), it is a popular and elegant ornamental tree, safe in cultivation in countries around the world, particularly subtropical regions. This tree also proves to be one of the hardier species, mature examples surviving perhaps -15°C, which makes it suitable for sheltered gardens in cool, temperate climates. Its juvenile leaves are quite dainty;

EUCALYPTUS PAUCIFLORA.

Opposite *Eucalyptus pauciflora* from *The Forest Flora of South Australia* by John Ednie Brown, 1882.

lanceolate, grey-blue and around 6cm long, held on stems with striking orange-red bark. Initially the plant is rather upright and quite bushy, but mature trees will develop appealing weeping branches, forming a wide, rounded rather dense crown. The narrow adult leaves are around 14cm long and look a little like those of willow (*Salix*), which explains the alternative common name of willow peppermint. The trunk has reddish-brown bark which peels away in fibrous strips when mature. Clusters of starry white autumn flowers held amid the evergreen foliage are a great bonus and look lovely spangling the cascading branches at a time of the year when many plants are fading.

In temperate gardens *Eucalyptus nicholii* is best given some shelter, in a hot, sunny location on well-drained soil. This is a good species to consider for a smaller garden because it will not rocket to the lofty heights of more often seen *E. globulus* or *E. gunnii*, for example. It makes a good specimen tree and will grace a gravel garden admirably, thriving and looking great with Mediterranean shrubs such as *Cistus*, *Cytisus*, *Phlomis* and *Laurus*. Underplant with shade-tolerant, drought-resistant species such as *Cyclamen hederifolium*, *Danae racemosa*, *Euphorbia amygdaloides* subsp. *robbiae*, *Geranium endressii*, *Hedera helix* or *Ruscus colchicus*.

OTHER *EUCALYPTUS* TO TRY

- ***Eucalyptus gunnii* 'Rengun' (France Bleu ®)** is a compact growing selection of cider gum, with impressive oval, bright blue juvenile foliage. It forms a bushy plant and can be kept as a coppice to maintain the best leaves. Height 3m. Hardy to -15°C.
- ***E. parvula*** (small leaved gum) has attractive small green-blue juvenile leaves, then longer silver-blue foliage. It is upright growing but stays fairly compact and is not too fast. Height 15m. One of the hardiest, -15°C.
- ***E. pauciflora*** is a small species from cold sites at altitude in eastern Australia. Desirable **subsp *niphophila*** is a slow-growing tree with bold leathery elliptical blue-green leaves and on mature trees marvellous white and grey bark. Height 8m. Hardy to -15°c.

Mrs Robb's bonnet

Euphorbia amygdaloides subsp. *robbiae*

Easy to grow, evergreen, ground-covering and suitable for sun or dry shade, this spring-flowering perennial is an invaluable plant to call on when conditions are tough.

Opposite Wood spurge (*Euphorbia amygdaloides*) from *English Botany, or Coloured Figures of British Plants* by James Sowerby, 1868.

This evergreen perennial is among the most useful of plants for its sheer adaptability, ease of cultivation, space-filling, ground-covering habit and, of course, long-lasting, vivid lime-green spring flower heads. It is one of few plants that will reliably flourish in dry shade under trees and shrubs, standing drought yet remaining leafy and attractive. Ordinary *Euphorbia amygdaloides* (wood spurge) is a fairly undistinguished clump-forming plant, native to European deciduous woodlands (including the UK) and east into Turkey. It has oval, initially rather hairy leaves held on stout shoots around 15cm tall, then in mid-spring these produce airy heads of bright green flowers – in all around 60cm tall, sometimes more, and spreading by seed. This plant is seldom found in gardens, although attractive selections such as 'Purpurea' with reddish foliage are often grown; these are sadly prone to succumb to mildew, especially as the going gets tough. A rather different proposition is *E. amygdaloides* subsp. *robbiae*. This is considered far superior as a garden plant by growers for its dark green, hairless foliage, held in neat, shining rosettes of growth atop stout stems. It has more vigour and a stronger constitution, suckering strongly by running roots and flourishing where little else survives. While Mrs Robb's bonnet does grow strongly, it generally stays just on the right side of being a nuisance: the suckers are easy to remove and, if lifted with a bit of root, can be potted up for new plants. It may also self-seed, so garden escapes into the wild are quite common and something to guard against.

This is a plant that looks good for much of the year – the dark rosettes are especially pleasing in winter. Plants are hardy to -20°C although long, hard spells of freezing weather can damage foliage. The spring flowers are produced by two-year old rosettes and last several months in beauty, developing reddish tints in sunnier spots. By late summer, however, it starts to look scruffy and this is the time to tidy up the clump, removing the flowered shoots at ground level because these will now have yellowing foliage. Do this job wearing gloves: this *Euphorbia*, like others in its tribe, oozes a toxic milky sap that can cause a rash on bare skin.

Opposite *Euphorbia amygdaloides* subsp. *amygdaloides* from *Herbier de la France* by *Pierre Bulliard*, 1780-98.

Few plants are as useful as this perennial – it's a plant to turn to when you know the conditions are challenging. Use it in shaded borders behind *Epimedium*, hellebores, Solomon's seal and various ferns. In brighter spots, thread through some orange tulips (such as 'Ballerina') or try *Allium hollandicum* 'Purple Sensation' in front – these saturated colours go well with the lime-green heads. Plant it around the base of a bamboo – a location that can be tricky – or even try it as a fringe in front of a conifer hedge. It looks great in drifts in a woodland garden – in those shallow, root-filled places where little else will flourish, alongside similarly tough *Vinca* and perhaps *Pachysandra*. Like these, Mrs Robb's bonnet is not a plant that stays put, and you'll have to tolerate new plants popping up in unexpected places such as paving cracks or along steps, but that's part of the appeal. Any colonies that don't please are easily whipped out. I recommend keeping it away from formal lawns where it easily invades, but in relaxed settings you might enjoy it popping up in tree circles, perhaps among daffodils and *Cyclamen hederifolium*.

You may be wondering who Mrs Robb was. A sharp-eyed botanist, horticulturalist and artist, Mary-Anne Robb was born in 1829 and we have her to thank for sneaking the plant into the UK. She'd admired this *Euphorbia* in woodland near Istanbul, and realized it would make a great garden plant, so she popped a piece into her hat box and brought it home.

OTHER *EUPHORBIA* TO TRY

- ***Euphorbia mellifera*** (honey spurge) is different. A shrubby, exotic-looking plant with handsome foliage to 3m or more, it bears yellow heads of honey-scented flowers in late spring. Needs full sun and sharp drainage, and is best with wall protection. Hardy to around -10°C.
- ***E. rigida*** makes a choice plant for a raised bed, gravelled bank, large rock garden. It has low, fleshy, sprawling stems bearing whorls of pointed grey-green leaves and green, often coral-tinged flower heads. It needs full sun and sharp drainage. 1m across, 40cm tall. Hardy to -10°C.
- ***E. myrsinites*** is a smaller version of ***E. rigida***, ideal for a rock garden or wall top with rounded blue-green leaves in whorls around fleshy sprawling stems terminated by round lime-green heads. It needs full sun and sharp drainage. 1m across, 20cm tall. Hardy to -15°C.

PLANTE VÉNÉNEUSE DE LA FRANCE. Pl. 95.

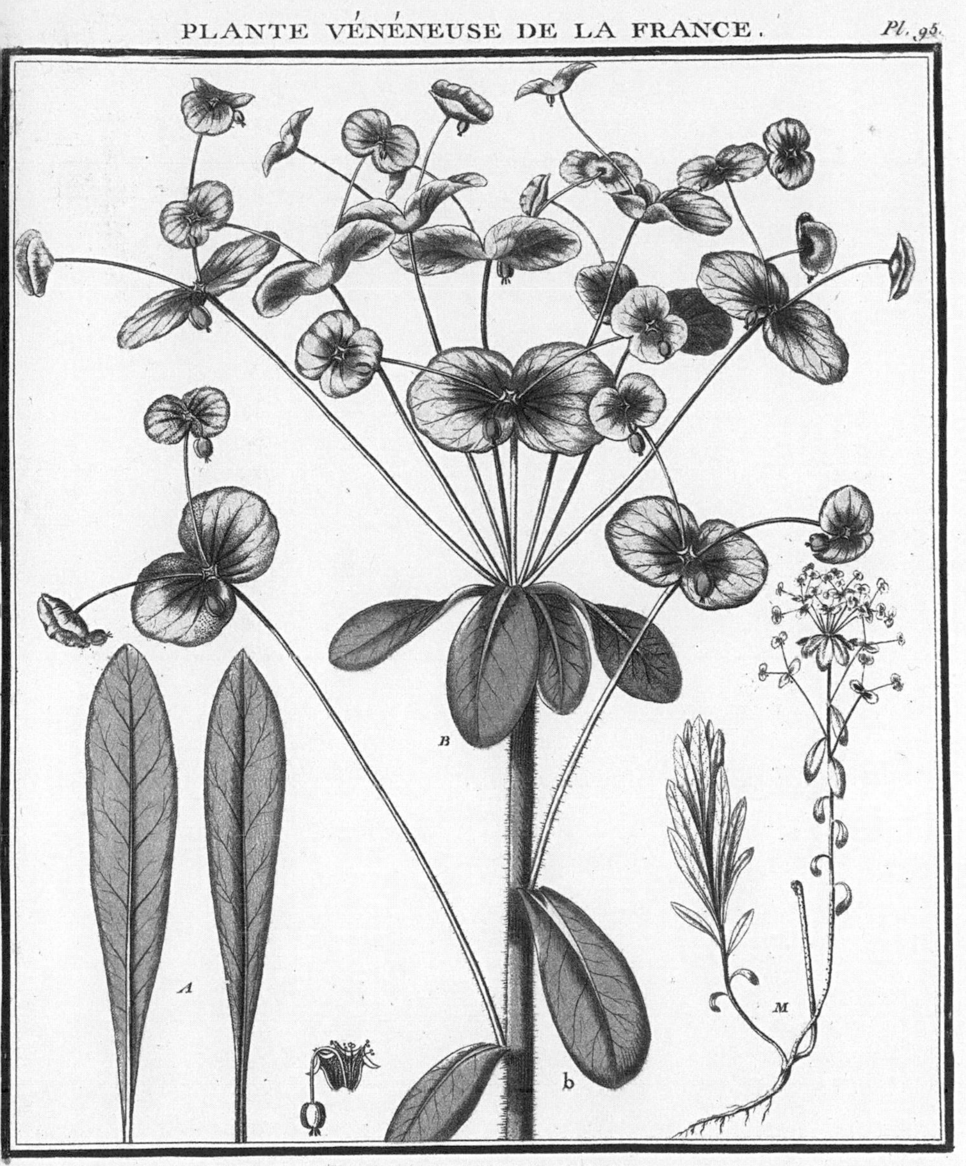

LE TITHYMALE DES BOIS. FLOR. FRANC.

Euphorbia sylvatica. *L. S. P. Dodec. Tryg. 663.* **PORT** *herbe bisannuelle qui fleurit en Juin, Juillet et Août, dans les bois.* **TIGES** *laiteuses, hautes de deux pieds ou environ, cylindriques, légérement velues. Ombelle à six ou huit rayons.* **FLEURS** *jaunâtres, composées d'un calice à quatre divisions très petites, de quatre pétales en forme de Croissant, de douze a dix-huit Etamines, et d'un Ovaire globuleux soutenu par un Pédicule assés long et surmonté de trois Styles bifides. Chaque Fleur a pour base deux Bractées orbiculaires, réunies et traversées du pédicule.* **FRUITS**; *capsules à trois Coques monospermes.* **FEUILLES**; *celles qui sont portées par les tiges de l'année sont lancéolées et ramassées en touffe à l'extremité superieure de la tige; celles des tiges fleuries sont obtuses.*

N. B. A. Feuilles des branches steriles., B. b. feuilles des branches fleuries. La fig. M. représente cette plante réduitte.

LE TITHYMALE DES BOIS, quand il est verd n'est ni moins âcre, ni moins dangereux que les autres especes.

Blue daisy

Felicia amelloides

This delightful low-growing daisy from South Africa makes a great drought-tolerant container plant in temperate gardens, producing masses of blue and yellow flowers; in sheltered places it may even survive winter outside.

Blue is a most unusual colour for a daisy, and this charming low-growing perennial produces multitudes of surprising flower heads that look almost like they were painted by a child: a yellow centre surrounded by contrasting, bright, sky-blue ray florets (the petal-like structures that fringe a daisy flower). *Felicia amelloides* is native to the southernmost parts of South Africa, especially coastal districts, where it is found – forming dense, evergreen ground cover, scrambling over old sand dunes, as well as dry, rocky slopes and hillsides – standing heat and dry periods and enjoying freely draining soils. The small elliptical leaves are rich green, rough and rather sticky to touch, held on brittle stems 30–50cm high. Flowers appear generously; they are small but striking, 3cm across held on slender green stems around 6–10cm above the foliage, attracting many pollinating insects. These flowers may develop into little dandelion-like clocks of seeds if plants are not deadheaded.

The display is most profuse in spring but may peter out through summer, especially in a hot season, to flush again when conditions cool. There are white or pale-blue flowered selections sometimes sold, but the common blue is most popular in gardens. *Felicia* is a largely African genus of around 80 species – this species is by far the best-known and has been grown in European gardens since the late eighteenth century. In the wild it may experience light frosts, but blue daisy is not regarded as hardy in most temperate climates. That said, it often survives increasingly mild winters in frost-prone regions, tolerating temperatures of around -2°C if plants are dry and drainage is sharp. In Mediterranean gardens it can be a winner, needing a little water only in really hot, dry periods and forming striking low ground cover in sun and softening the edges of rocks walls and paving, perhaps forming a mound around the bases of succulent *Agave* and *Aloe*, beside multi-hued *Lantana camara* or in front of rosemary, lavender or oleander (*Nerium*). Plants look and flower best if you can regularly deadhead them – simply trim off old flowers complete with stems with a pair of snips or scissors to the level of the foliage. This also helps to keep plants compact and tidy, especially important when the plant is part of a container display.

Opposite Blue daisy (*Felicia amelloides*) from *L'Illustration horticole*, 1861.

P. Bessa pinx.

Barrois sculp.

Cineraria amelloidea

Opposite Blue daisy (*Felicia amelloides*) from *Herbier général de l'amateur* by Jean Claude Michel Mordant de Launay and Jean-Louis-Auguste Loiseleur-Deslongchamps, 1824.

In cooler climates, *Felicia amelloides* is widely grown as summer bedding for use in containers – they are useful in this respect because they need far less regular watering than the usual petunias and begonias. These compact daisies look particularly good planted on their own in wide, shallow pots which allow plants plenty of space to form a mat of blue flowers – containers can then be placed in front of other planted pots to create a stepped summer display. They also fit well in wall pots and window boxes, perhaps alongside other sun-loving, fairly drought-tolerant choices such as pelargoniums and succulent *Delosperma* and *Carpobrotus*. In winter you can move plants into a frost-free glasshouse; they will look good for two to three years using this method. The alternative is to take cuttings of stem tips, either in late summer for growing on a sunny windowsill or in a greenhouse over winter, or in spring, from old plants kept under cover. If you have a really sunny wall top, rock garden or a sheltered corner, it is also worth experimenting with planting *Felicia* outside permanently in a sheltered pocket. Grow it beside other low-growing, sun-loving plants that like sharp drainage, such as similarly daisy-flowered *Erigeron*, dainty heron's bills (*Erodium*), dazzling rock roses (*Helianthemum*), marvellous yellow-flowered *Hypericum olympicum*, various succulent carpeting *Sedum* and mounds of houseleeks (*Sempervivum*). It is worth looking out for variegated selections such as *Felicia amelloides* 'Santa Anita', which has cream-edged foliage that makes a great foil for the flowers.

ANOTHER *FELICIA* TO TRY

- ***Felicia petiolata*** is a close relative of blue daisy, bearing masses of pinkish-mauve flowers all summer and autumn on long, reddish, spreading or scrambling stems; they weave through other plants, even up through hedges and other plants, or cascade down a sunny bank or over a wall. It is a much hardier plant than ***F. amelloides***, surviving at least -10°C, and also stands drought well, but is usually rather less tidy in appearance.

Giant fennel

Ferula communis

A spectacular, tall-growing herbaceous perennial with mounds of feathery growth early in the season followed by towering stems bearing heads of bright green flowers.

This imposing umbel (member of the cow parsley family, Apiacae) is a native of the Mediterranean and east Africa, growing in scrubland and open, rocky, rather grassy places, thriving in unrelenting heat and enduring parched summers; plants are often admired from passing airport taxis erupting into bloom on Italian and Greek roadsides, when plants develop soaring, branched flower stems 4–5m tall and bearing thousands of acid yellow-green flowers, held in numerous spherical heads – a wonderful sight. *Ferula* are close relatives of common fennel (*Foeniculum*) but larger and more spectacular, and with little aniseed scent when foliage is crushed. There are around 200 species, some growing across into Asia, but few are met in gardens. Some are used for their aromatic resins; *F. gummosa* is the source of galbanum, used in ancient Egyptian times as incense and occasionally a component of modern perfumes. *Ferula* are herbaceous perennials, some monocarpic, the parent plant dying after setting seed. Happily for gardeners, *F. communis*, the species most often cultivated, is usually long-lived. Its appeal as a garden plant begins early in the year. As soon as January, foliage begins to arise. This is delicate and finely divided, the feathery leaves ever larger, rich-green and shining, yet also surprisingly resistant to cold and frost – plants are hardy to around -10°C. It is, in early spring, perhaps the freshest-looking of all garden plants, a real taste of the season to come and providing much-needed structure and interest. By mid-spring plants develop into a fabulous, frothy, shimmering mound around 1m tall (more across), a perfect verdant backdrop to tulips and other bulbs. It is now that flowering stems may develop – these are stout and succulent, often purple-flushed and carrying wide-based leaves that sheath stems.

Opposite Giant fennel (*Ferula communis*) from *Curtis's Botanical Magazine*, 1907.

Flower heads develop and open before the plant reaches its full height, usually in June. After flowering, the plant gradually dies back. In young plants, removing the stem may help it be longer-lived, but when plants are established it is worth allowing them to set seed – this can be collected when ripe in autumn and sown; seedlings will arise freely.

Ferula communis take a few years from seed to reach flowering size, and even then will not always bloom annually,

Opposite Giant fennel (*Ferula communis*) from *Flora Graeca* by John Sibthrop and James Edward Smith, 1819.

the plant taking time off to build strength. Some growers report that this habit can be avoided by mulching plants with well-rotted manure, and established plants will develop numerous flowering stems. When dry, these are surprisingly enduring – hollow, still-standing remains are often seen in Mediterranean landscapes months later; the Romans are said to have used them as lightweight walking sticks.

In gardens these plants like well-drained, fertile soil in full sun, with plenty of space. Remember they get large, and as plants form deep taproots so they cannot be moved once established. Plant them as young as possible – seedling stage is best, larger potted plants are likely to be short-lived. Giant fennels are ideal for gravel gardens, where they make an arresting sight with their architectural form, perhaps beside *Yucca*, larger *Eryngium* and sun-loving shrubs such as *Cistus* and rosemary. Alternatively try them at the back of a border and plant tulips and alliums, honesty (*Lunaria*), *Hesperis* and oriental poppies (*Papaver orientale*) in front. Giant fennel's only fault is that its foliage yellows before summer's end, leaving a gap, a shortcoming resolved by masking the site with plants that look good late in the season, such as asters, *Miscanthus*, *Helianthus* and *Verbena bonariensis*.

OTHER UMBELS TO TRY

- ***Bupleurum fruticosum*** is a shrub. Evergreen and drought-tolerant, it loves sun and provides masses of green flowers in spring and early summer. Height 1.5m. Hardy to -10°C.
- ***Ferula tingitana*** is another giant fennel sometimes grown. Smaller, with less finely cut foliage, it may be easier to fit into gardens. Height 2m. Hardy to -10°C.
- ***Myrrhis odorata*** (sweet cicely) likes shade, developing cow parsley-like flowers in spring and ferny foliage. In a dry season it dies down early, surviving drought. Height 1.2m. Hardy to -15°C.

Common fig

Ficus carica

Cultivated since ancient times in the Middle East, this handsome, productive tree is almost heat- and drought-proof. Many selections are grown in gardens around the world, bearing green, yellow or purple fruits with a delicious honey-sweet flavour.

Opposite Common fig (*Ficus carica*) from *Plantae Selectae* by Christoph Jacob Trew, 1771.

Many long cultivated, drought-tolerant plants come from the Middle East and as a result, among the earliest mentions of them are from the Bible; figs are one example. Today these sun-loving plants are grown for their sweet, distinctively crunchy, succulent fruit and the beauty of their lobed leaves, thriving in poor soil that is often parched in summer. Native to the Mediterranean and Asia, common fig can be found in the wild sprouting from cracks in bare rockwork, their questing roots searching out sources of water. Fig trees are deciduous and reach 10m tall, often rather more across, with a broad dome-shaped canopy and smooth, pale grey trunk in time. They are important shade trees, providing cooling shelter through the heat of the day. The rough green leaves can reach 40cm or more long, and may be whole and heart-shaped or more typically hand-shaped with three to five finger-like lobes. These leaves are also pleasingly aromatic, the scent from them strongest through the hottest, sunniest hours.

Figs are among the longest cultivated of plants: the fruits were certainly eaten in Neolithic times, and archaeological finds from the Middle East suggest they may have been actively planted and grown around a thousand years before wheat, in what may be the earliest form of agriculture. The Romans knew at least 30 different selections, at least one of which is likely to have been *F. carica* 'Dottato', which is still available today. In many countries figs are sun dried to preserve them through winter.

An individual fig is not strictly a fruit but a synconium, a hollow, hot air balloon-shaped structure, the inside of which is lined with minute flowers. These are pollinated by a little wasp which enters the fig through a hole at one end. After pollination, the flowers develop into tiny one-seeded fruit within the fig, and the structure swells and ripens, hanging irresistibly from the tree. Little is tastier than a juicy, sun-warmed fig and its inevitable consumption aids the plant's seed dispersal. Testament to this are the numerous fig trees that still flourish on the banks of many English rivers, such as the Thames and the Don in Sheffield. In Victorian times figs were popular fruits and these rivers badly polluted by sewage.

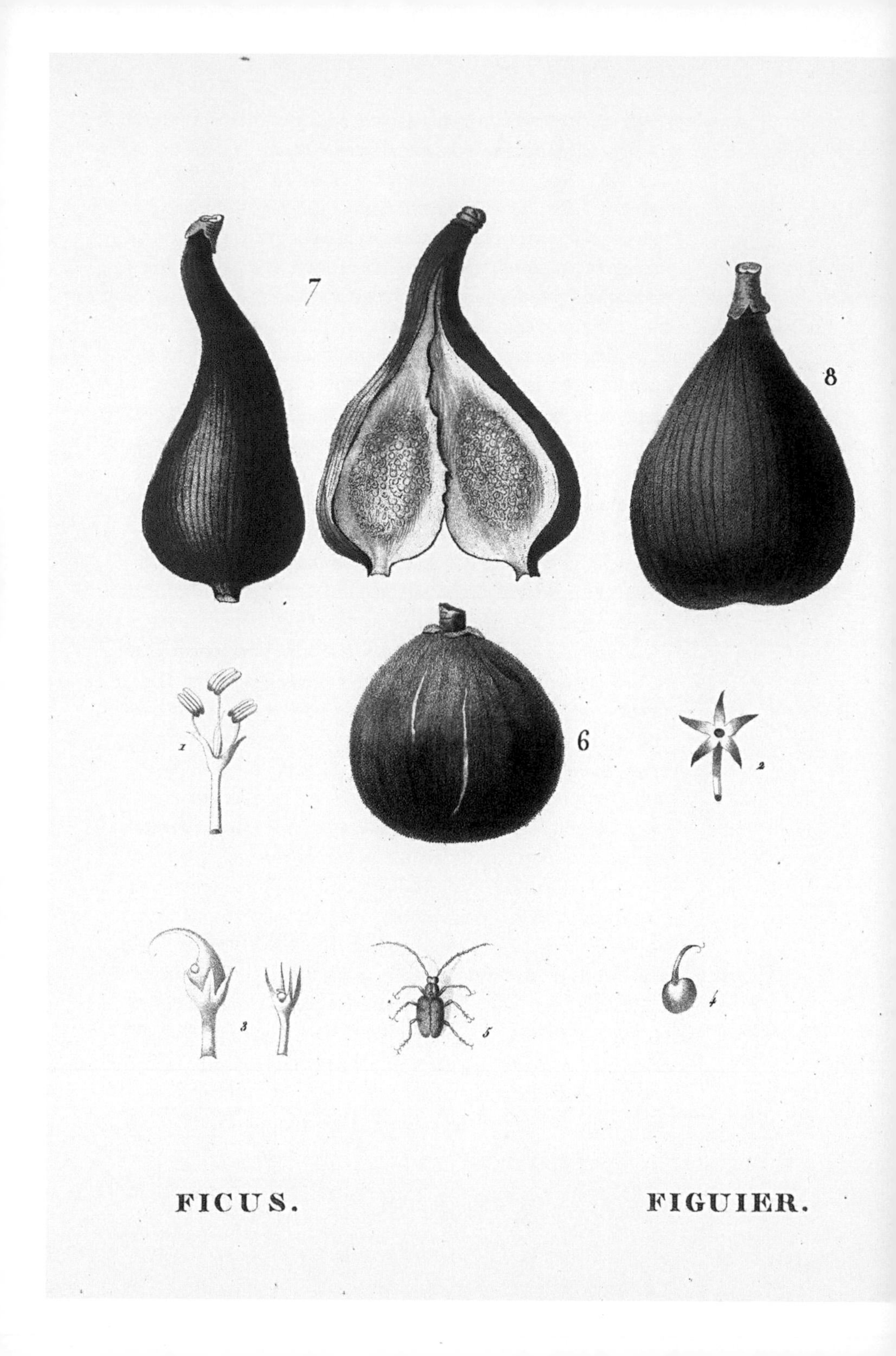
7
8
6
1
2
3
5
4
FICUS.
FIGUIER.

Opposite Common fig (*Ficus carica*) from *Traité des arbres fruitiers* by Duhamel du Monceau, 1809.

In gardens, fig trees have so much to offer, needing little care and standing drought remarkably well once established. They also lend a distinctly Mediterranean feel to plantings, even in frost-prone regions, although for good crops of fruit in cool, temperate areas some cultivars prove more reliable than others. In warm climates they flourish and fruit easily as free-standing trees, often providing two crops of fruit per year. The early crop is produced from figs that developed on trees at the end of the previous year and go on to develop fully the following spring. These early fruits are a highly prized delicacy in southern Italy, where they are known as *fioroni*. In colder regions the little embryo figs do not survive winter, so gardeners in more northerly climes rely on the second, heavier crop in late summer and autumn. To get figs to ripen before the end of summer in these cooler regions, plants are often trained flat on to sunny, south-facing walls where they can make beautiful garden features. Restricting the roots with bricks lining the planting pit aids fruiting. The most widely planted of these tough selections are 'Brown Turkey' and 'Brunswick' but others are worth a try.

Many Victorian walled gardens grew figs wall-trained under glass in purpose-built fig houses, which allowed the cultivation of some of the most delicious figs of all, such as sensational 'Rouge de Bordeaux' or 'Violette Dauphine' which need winter protection and summer heat to ripen. Today you can enjoy these selections by growing plants trained as standards in large containers kept in a glasshouse – they are easy to keep grown in pots and crop well.

If you have the need to prune your fig, beware the milky sap from cut stems: it is a skin irritant, so wear gloves. This is a job best done when plants are leafless in late winter or early spring.

MORE HARDY FIGS FOR A SUNNY WALL

- **'Osborn's Prolific'** produces heavy crops of tasty purple figs.
- **'Panachee'** bears sweet, yellow striped, green figs.
- **'White Marseilles'** has green fruit with sweet white flesh.

Mount Etna broom

Genista aetnensis

This hardy, late summer-flowering shrub or small weeping tree is ideal for hot, dry sites, bearing masses of golden-yellow, scented flowers.

A magnificent flowering shrub or small tree when mature, Mount Etna broom makes a sensational sight seen at its best, with a rounded weeping crown of slender, rich green branches laden with sweetly scented, golden-yellow, pea-shaped flowers from mid to late summer, which make a beautiful contrast to azure-blue Mediterranean skies. As its name hints, it is native to the island of Sicily (and also Sardinia) and naturalized in many other areas, and does indeed grow on the lower sun-baked open slopes of the titular volcano in poor, well-drained soil. Like many other brooms, its rush-like branches bear only slender, short-lived leaves when the branches first develop. As these mature, the leaves fall off, the green branches carrying out the essential task of photosynthesis. Pea-like seed pods follow the flowers, and all parts of the plant are toxic.

At first *Genista aetnensis* is a bit of an ugly duckling. Although they will flower, young plants are not much to look at – shrubby, generally upright at first but rather awkward and shapeless with an open habit. After around six to eight years, however, they gradually become more tree-like, plants shedding lower branches to reveal either a single trunk or a multi-stemmed form, the upper branches reaching up to provide a domed canopy, arching out and then cascading over. The plant will reach around 8m tall and often more across, becoming a somewhat irregular but highly characterful, graceful tree. Plants prove fairly hardy, surviving lows of around -15°C when established.

In gardens they enjoy an open site best, although in cooler, temperate regions, the reflected heat from a nearby sunny wall is much appreciated. Mount Etna broom grows well in exposure and can thrive in maritime areas, but if planting in such a position, always choose a young plant that can get its roots down into the soil. The older the plant, the greater the chance of wind rock, which will prevent the plant properly establishing. This fact helps explain why *Genista aetnensis* remains a seldom-seen plant: young plants simply do not appeal much at garden centres. A shame because a little patience reveals that this plant has many virtues, impressive drought resistance and ease of cultivation high among them.

Opposite Mount Etna broom (*Genista aetnensis*) from *The Garden*, 1893.

Opposite Mount Etna broom (*Genista aetnensis*) from *Curtis's Botanical Magazine*, 1826.

In a gravel garden, this broom will become an impressive living focal point in time, flowering late in the summer when little else is at its best. Many drought-tolerant Mediterranean shrubs are quite quick to grow and mature, and may have short lifespans of no more than 10 years in a garden. This plant, however, is in it for the long-term and is best considered a key plant from the outset. It needs little attention – the tree-like form can be encouraged by the removal of the lowest branchlets and training the main stem on to a tree stake, but mature plants do not need any pruning; plants will not shoot if cut back into old wood.

When planting, try contrasting it with the upright, slender forms of *Cupressus sempervirens*, *Juniperus scopulorum* 'Skyrocket' or even giant reed *Arundo donax*. Plant other lofty grasses such as *Cortaderia* or *Ampelodesmos mauritanicus* in front, while that usefully open, airy crown means you can easily underplant too; the yellow flowers look superb cascading on to bold blue *Agapanthus*, airy *Salvia yangii* or *Hibiscus syriacus* 'Oiseau Bleu'. It will also look effective at the back of a mixed border, as a specimen in the centre of an island bed or in colder areas planted beside a sheltered house wall weeping over a sunny terrace.

OTHER BROOMS TO TRY

- ***Genista lydia*** (**Lydian broom**) is a super low-growing deciduous shrub with ground-covering growth that is plastered with yellow flowers in early summer, great for dry, sunny banks and gravel gardens. 50cm tall, 1m spread. Hardy to -15°C.
- ***G. pilosa*** **'Procumbens'** (creeping broom) makes a great mound-forming choice for a sunny rock garden, with creeping stems bearing dark-green leaves and little golden-yellow flowers in summer. 20cm tall, 50cm spread. Hardy to -15°C.
- ***Retama monosperma*** (bridal veil) is a white-flowered, bushy or tree-like broom, highly drought-resistant with upright, then arching silvery-green stems that are covered in blossom in spring. Needs a sheltered site in temperate gardens. 3m tall, 3m spread. Hardy to -5°C.

Knotted cranesbill

Geranium nodosum

Easy, hardy, low-growing perennial great for shade and surviving spells of drought well, producing mauve-pink or occasionally white or purple flowers through summer and autumn.

Obliging is a highly suitable word to sum up this superb summer-flowering perennial, for there are few hardy plants more adaptable for flourishing in all manner of positions around the garden, including, most helpfully, dry and quite deep shade. While there are many larger-growing, more extravagant geraniums (cranesbills), this species and its cultivars should be towards the top of any gardener's wish list.

A European native found in woodlands through mountainous parts of France and Italy including the Alps and Pyrenees, *Geranium nodosum* withstands periods of summer heat and drought in the wild. As a genus, *Geranium* is widely admired by gardeners, most species easy to grow and with a long flowering period; this plant is no exception. Knotted cranesbill forms soft, rather sprawling mounds of three- or five-lobed, bright green leaves which emerge in spring from rhizomes that spread out just below soil surface. Each leaf has a slightly puckered, shiny surface and collectively they form an appealing, relatively slowly spreading low mound of growth. Plants are normally no more than 30cm high and as much across. The five-petalled flowers arise in early summer, opening to 2.5cm across, and are commonly shimmering lilac-pink in hue, with a pale centre to each funnel-shaped bloom. Petals usually feature contrasting dark veins, while tips may be irregularly lobed, the blooms held in sprays on slender, almost wiry, reddish stems so that they seem to almost float above the leaves. Flowers are followed by pointed, explosive seed capsules typical of cranesbills, which pop open when ripe to scatter the contents. It is a hardy perennial, surviving -15°C.

In gardens, *Geranium nodosum* makes excellent carpeting ground cover, planted either en masse or with other plants, perhaps tougher ferns, *Epimedium*, *Helleborus*, *Liriope*, *Pachysandra* and *Polygonatum*. Its sprawling flower stems and free-seeding habit are ideal for knitting plants together to provide a naturalistic, free-flowing feel, forming a veritable tapestry of cover below trees and shrubs. You can also include plants towards the front of mixed or herbaceous borders, in shaded corners in a gravel garden, even under roses. The flowers

Opposite Knotted cranesbill (*Geranium nodosum*) from *English Botany, or Coloured Figures of British Plants* by James Sowerby, 1864.

Geranium
Macrorhizon.

Opposite *Geranium macrorrhizum* from William Curtis Collection, 1787-99. Kew Collection.

are produced over a long season – the main flush early on but then sporadically into autumn, individual blooms popping open here and there. It is one of few plants worth trying below evergreens, even at the outer edge of the canopies of many conifers. This geranium does a rather good job at covering fading foliage of many flowering bulbs such as snowdrops and daffodils because its clumps expand through the progress of spring. Usually, the plant needs no special care or watering; in hot, sunny places, clumps can look shabby around midsummer, in which case you can take a pair of shears and cut the lot back to ground level. Give the plant some water, and it will regrow and reflower for the end of summer. The leaves are often green through most of winter. Plants are easily lifted and divided in autumn for bulking up plantings; they also spread by seed, young plants springing up in unexpected places. Seedlings may vary from the parent and it's worth looking for distinct plants to keep.

There are numerous selections to collect to fill in tricky spots in dry shade around the garden. Outstanding *G. nodosum* 'Silverwood' has white flowers that shine beautifully in shade above fresh green foliage – a first-rate plant; 'Whiteleaf' bears masses of pink-purple flowers; and 'Blueberry Ice' has lilac-purple flowers, each petal edged in pale pink with contrasting dark purple veins. For lovers of jewel-like flower colours, both 'Clos du Coudray' and 'Julie's Velvet' have shimmering velvet-purple flowers edged with white that sparkle amid foliage.

OTHER *GERANIUM* TO TRY

- ***Geranium endressii*** is a tough, long-flowering mound-forming perennial with pink flowers, for sun or part-shade, making good ground cover. Self-seeds freely. Cut down in midsummer if conditions are too hot and dry; it will re-sprout. Height 40–50cm, -10°C.
- ***G. macrorrhizum*** is low-growing with sticky, aromatic foliage forming a dense mat of leaves in sun or shade, and standing drought well. Bears pink flowers in spring held on low stalks and develops reddish autumnal tints. Height 30cm, -10°C.
- ***G. × magnificum*** flowers in late spring, bearing masses of superb bright purple-blue flowers above its rather hairy leaves. In hot sites choose a place in part-shade. Height 50–60cm, -10°C.

Chapparal yucca

Hesperoyucca whipplei

Sun-loving, architectural plant forming impressive rosettes of slender blue-green leaves and eventually producing soaring flower spikes bearing hundreds of individual blooms.

If you are looking for a potentially spectacular plant to grow in a hot, dry position in full sun, this remarkable *Yucca* relative may be the perfect plant to choose. Yuccas and their near relatives, including *Hesperoyucca*, are tricky plants to categorize, some authors referring to them as herbaceous perennials, others regarding them being closer to shrubs. Many species, including this one more closely resemble succulent – and related – *Agave*.

Hesperoyucca whipplei is native to southern California and Mexico, found growing in chapparal plant communities on sunny arid rocky slopes, often in mountainous areas, and even as part of oak woodland. For several years the plant slowly builds a roughly spherical rosette, composed of hundreds of slender, rigid, silvery-blue, rapier-like leaves, 20–90cm long, each one with a finely serrated margin and tipped with a sharp black spine. Individual plants eventually reach up to 1.5m across. In this species no trunk is formed.

Eventually, once it has banked enough energy, Chaparral yucca will flower. Exactly how long this takes to come about varies depending on various factors, but it will be at least five years in the wild, rather longer grown in cool, temperate gardens, but it is a truly extraordinary thing to witness. From the centre of the plant arises an erect, sturdy green stem, at least as thick as a broom handle, quickly soaring to around 3m tall. From the top half of this develop dozens of radiating side branches, 30cm or more long, each holding masses of purple-tinted, ivory-white, bell-shaped, scented flowers – the whole head in all producing many hundreds of blooms. This plant's alternative common name, our Lord's candle, is highly apt: when in bloom it resembles a gigantic candle in a holder. Flowers in the wild are pollinated by a particular kind of moth which also lays its own eggs when fertilizing the blooms, the offspring later feasting on the plant's seeds. Individual rosettes usually die off after flowering, but not before offsets have formed at the base of the plant. The faded flower spikes endure long in the landscape.

The plant is widely used as an ornamental in California and will thrive well in any dry Mediterranean climate, flourishing on dry slopes and terraces alongside similarly formed *Agave*

Opposite Chapparal yucca (*Hesperoyucca whipplei*) from *Curtis's Botanical Magazine*, 1899.

Opposite Chapparal yucca (*Hesperoyucca whipplei*) from *Revue horticole*, 1884.

and other succulents. It was introduced to Europe in the middle of the nineteenth century but is seldom grown in cool, temperate regions because it has never been regarded as easy to keep, unless you can provide sharply drained soil and that all important hot, sunny site. Having said that, plants prove hardy to around -10°C if kept fairly dry, so in increasingly hot, dry summers and often milder winters due to our changing climate, they are well worth trying, particularly in urban areas. An alternative is to grow it in a large container on a sunny terrace for overwintering under glass, kept dry. When they form, the flowers usually do so in late spring, which means there is time for the spike to fully develop before the first frosts of autumn, a failing of some more widely grown *Yucca*. Pleasingly attractive as foliage plant, with its sculptural form and arresting leaf colour, *Hesperoyucca whipplei* looks great planted in a gravelled area, perhaps with *Beschorneria* and palms such as *Chamaerops* or near softer drought-resistant plants including *Artemisia*, *Cistus* and *Epilobium canum*. You can even sow *Eschscholzia californica* at its feet – certainly it is best not to plant it anywhere too close to paths, where its sharp leaf tips may cause damage. Raised beds make an appropriate spot, especially if they benefit from the reflected heat and rain shadow effect of a nearby wall, while maritime gardens are another possible location to consider.

OTHER *YUCCA* TO TRY

- ***Yucca filamentosa*** (Adam's needle) bears stemless rosettes 40–60cm wide of spine-free, lance-shaped leaves, the edges of which bear fine white threads. Showy flower spikes arise in summer. 1.5m, 15°C.
- ***Y. gloriosa*** (Spanish dagger) is the best-known yucca and an easy to grow, hardy plant with rosettes of broad, stiff, spine-tipped lance-shaped leaves and eventually a short trunk. Impressive flower spikes, 2m tall, arise in late summer, running the risk of frost damage. 2m, -15°C.
- ***Y. rostrata*** makes a great alternative to ***Hesperoyucca whipplei*** for a hot, dry site and, when young, is superficially similar with slender blue-green, spine-tipped leaves. In time this plant forms a trunk, becoming tree-like. The still-impressive flower spikes are shorter. 3m, -15°C.

Yucca Whipplei violacea.

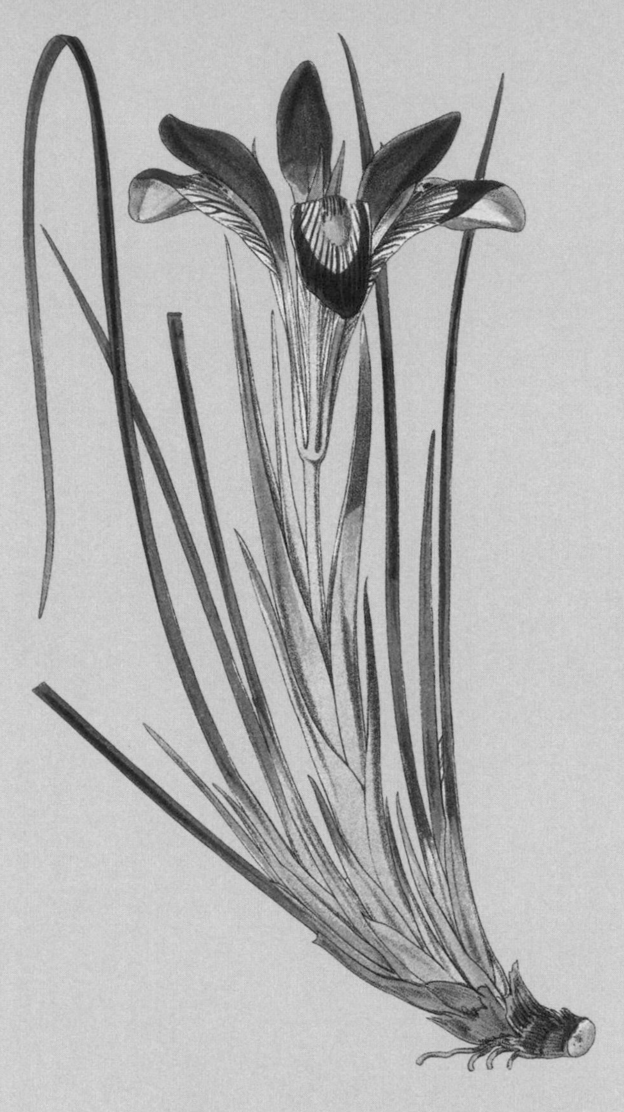

Algerian iris

Iris unguicularis

Low-growing, sun-loving hardy perennial standing summer drought with ease and which all through winter bears a succession of delicate mauve or white, sweetly scented flowers.

inter-flowering plants are always cherished by gardeners in cool, temperate regions; those that can also stand long spells of summer heat and drought with minimal care are few and far between, something that makes this little iris rather precious.

Opposite Algerian iris (*Iris unguicularis*) from *Curtis's Botanical Magazine*, 1878.

Native to the Mediterranean, particularly Greece, Crete, Turkey and parts of north Africa, it is a clump-forming, evergreen perennial, with long, rather grassy, upright and then arching leaves to around 60cm or so, developing from small rhizomes that grow just below the soil – it is these that help it endure long, dry, baking hot summers, growing on rocky terrain in poor soil. The showy flowers appear in winter when conditions are cooler and wetter, and are held just above soil level within the foliage, on a short stem (which is in fact the flower's style). Plants are rather variable, but the delicate flowers are usually shimmering satin shades of mauve, lavender or purple, with yellow markings on the broader lower petals (known by iris growers as falls) and with erect, rather small upper ones (standards). The flowers have a sweet perfume. Individual blooms last just two or three days but they spring up for months on end – if frost or foul weather ruins one flush, another follows, all the way into spring. After the flowers, seed pods form – these often go unnoticed, hidden at the very base of the plant, but it's worth catching these before they split open and sowing a few seeds because they soon make flowering-sized plants.

In gardens *Iris unguicularis* are much appreciated for winter colour and can be used in winter gardens, but remember that to flower well they do need that summer bake. It is often said they like really poor, almost rubble-like soil – and certainly they will survive in such a site – but ordinary garden soil that is well drained will probably give better results. What they need most is an open site; these are not plants that appreciate too much competition around them, although they do fine with spring and summer bulbs – crocus, smaller daffodils and alliums perhaps, and smaller airy annuals such as *Nigella* and *Eschscholzia*.

Plant this iris at the foot of a sunny wall, in a rock garden or even in a gravel garden, but ideally you'll want to find a sunny, slightly out-of-the-way place – in truth, plants

Iris Stylosa.

Opposite Algerian iris (*Iris unguicularis*) from *Revue horticole*, 1900.

can be pretty unsightly through summer. Although the plant is evergreen, the end of each leaf tends to turn brown and clumps need to be persuaded to part with yellowing leaves – an occasional comb through helps keep them looking presentable, but by late summer it can be best to trim the foliage in height a little and carefully pull out the worst of the thatch if the plant is getting tatty. Plants start into growth in autumn, fresh new leaves then emerging followed by the flowers. This is also the time to divide old clumps – new roots will also be produced now. Splitting clumps should be done only occasionally – they are not quick-growing and dislike disturbance, but after a few years they do tend to go rather bare in the middle, the new vigorous shoots produced at the outer edge.

There are quite a few named selections of *Iris unguicularis* worth seeking out; 'Alba' makes a lovely contrast with its white and yellow flowers but is not especially strong-growing and wants a bit of care; 'Mary Barnard' is more vigorous and bears violet purple flowers. Little 'Peloponnese Snow' is gorgeous with purple and yellow marked white falls and best in a rock garden. The much larger 'Walter Butt' has big pale lavender-blue flowers – amid copious tall foliage – and these blooms can be top-heavy, so are best picked and popped in a bud vase.

OTHER IRISES TO TRY

- ***Iris florentina*** is a long-cultivated, white-flowered bearded iris with broad, upright foliage, in all much larger than ***I. unguicularis*** and flowering in late spring, but another sun lover and also drought-tolerant. Grows to 60cm and hardy to -20°C.
- ***I. foetidissima*** (roast-beef plant) is a clump-forming European native and thrives in dry shade. Its yellow or purple summer flowers are followed by fruit that split open in winter to reveal showy, usually orange seeds. Grows to 50cm, hardy to -20°C.
- ***I. innominata*** (Pacific Coast iris) forms low clumps of slender grassy foliage in sun and well-drained soil, and stands dry summers well. The showy late spring flowers vary greatly in colour; there are many fine hybrids available. Grows to 20cm, hardy to -15°C.
- ***I. pallida*** (Dalmatian iris) has silvery-green foliage and in late spring, scented pale lavender flowers held on tall stems. Likes sun and dry soil. Grows to 90cm, hardy to -20°C.

English lavender

Lavandula angustifolia

Long grown for their floral beauty, as well as aromatic and medicinal qualities, lavenders thrive in sun and poor, well-drained soil, and adapt to a wide range of planting styles.

England is not a place often associated with a lack of rain, so it surprises many to learn that parts of the country, particularly the east, have lower rainfall than the national averages of Greece or Italy. As a result, gardeners in these regions – and further afield – have long turned to drought-tolerant plants to fill summer borders with flower, and lavenders – particularly *Lavandula angustifolia* – are among the best-known. This much-loved, easy-to-grow plant is synonymous with classic, 'English-style' gardens, which explains the common name, but like so many other national favourites, this is a plant simply adopted by the English and has subsequently provided many of the named lavender selections popular in gardens, such as widely grown 'Hidcote', 'Imperial Gem' or 'Munstead'. It's actually a Mediterranean native, found growing wild in parts of Spain, Italy and Croatia, among other countries, although other species are also found through the Middle East and into Africa.

Lavandula angustifolia was used widely by the Romans for its strongly aromatic foliage which repels insect pests, and they likely spread the plant to the furthermost reaches of their empire. Certainly, it was in widespread cultivation in England by Tudor times and used for a wide range of purposes from deterring bed bugs to mixing with beeswax for furniture polish. Today it is grown in countries around the world, with cultivars of this species cultivated commercially for the production of lavender oil, notably in Norfolk and Provence in the south of France. It is a compact-growing, dense, twiggy rounded shrub to 2m high (usually no more than 1m) and much the same across in flower; the narrow, lance-shaped evergreen leaves are a silvery grey-green, strongly aromatic and up to 6cm long. In summer, masses of flowering spikes emerge, held on long, very slender, leafless stems. The purple individual flowers are small and tubular and the source of lavender oil, and clustered densely at the end of the stems, the display lasting in beauty for many weeks.

Opposite English lavender (*Lavandula angustifolia*) from *Traité des arbres et arbustes* by Duhamel du Monceau, 1806.

Plants prove hardy to around -15°C, especially when grown in a well-drained site. In gardens it proves a highly versatile plant, suited to a range of uses and places, but thriving best in free-draining soil somewhere in full hot sun. Once established, it is impressively impervious to summer drought

Opposite English lavender (*Lavandula angustifolia*) from *Köhler's Medizinal Pflanzen* by Franz Eugen Köhler, 1887.

and has long been used to form low step-over hedges in more formal planting schemes, making a good alternative to box hedges for edging pathways or simple knot gardens. Individually plants can be dotted through sunnier borders, where they will form relaxed domes of flower in time, while that evergreen habit allows lavenders to be structurally significant through winter. They look super lining a broad path alternated with clipped box or balls of Japanese holly (*Ilex crenata*), perhaps with a scattering of *Allium hollandicum* 'Purple Sensation' threaded through; they also work well in front of roses, a classic pairing, as long as the spacing is not too tight. Use them to add flower power to herb gardens, or plant in containers on a sunny terrace – they are good candidates for use in a pair of matching pots either side of a doorway or flight of steps. And wherever you plant them, the insects will follow – honey and bumblebees especially but also hoverflies and butterflies are attracted.

Plants are not long-lived unless they are regularly pruned. It is best done in late summer – when conditions are dry and warm. Simply remove the top third of stems, cutting with sharp shears or secateurs just above next year's shoots which will just be visible. This keeps growth tight and shapely, plants lasting 10-15 years, if not longer.

OTHER LAVENDER TO TRY

- ***L. angustifolia*** '**Ashdown Forest**' is a first-rate selection with glowing purple flowers that begin early in summer and have great perfume. Height 70cm.
- ***L. angustifolia*** '**Purity**' is one to try for white flowers which contrast well with other classic lavenders – compact-growing, it reaches around 60cm high. '**Loddon Pink**' has pale rose flowers and reaches 50cm.
- ***L. × intermedia*** is a hybrid between ***L. angustifolia*** and ***L. latifolia.*** Look out for cultivars such as '**Olympia**' ('**Downoly**') 75cm tall with really dark purple-blue flowers; '**Grosso**' which is used in oil production and reaches 1m; or cream variegated, 70cm '**Walberton's Silver Edge**' ('**Walvera**').
- ***L. stoechas*** (French lavender) is a more tender species (surviving perhaps -7°C) with dark purple, oval flower heads topped off with ear-like mauve bracts – there are many selections to choose. It likes sharp drainage and full sun. 1m.

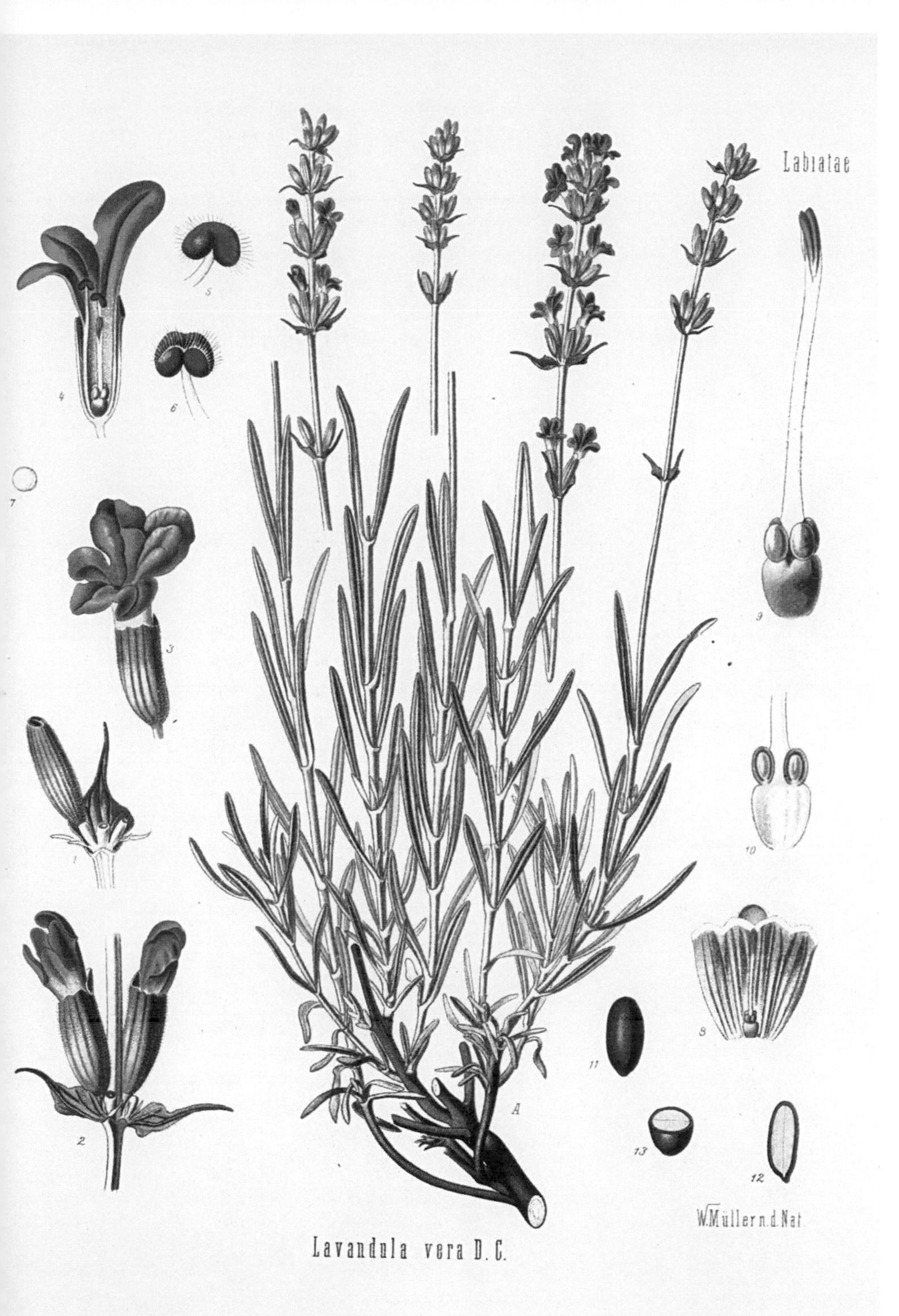
Labiatae
4
5
6
7
3
1
2
9
10
8
11
13
12
A
W. Müller n. d. Nat.
Lavandula vera D. C.

Lion's tail

Leonotis leonurus

An unusual plant that forms a shrub in warm climates but is often treated as an annual in temperate areas, and bears impressive whorls of orange flowers atop tall stems.

Opposite Lion's tail (*Leonotis leonurus*) from *Herbier général de l'amateur* by Jean Claude Michel Mordant de Launay and Jean-Louis-Auguste Loiseleur-Deslongchamps, 1824.

This arresting, upright shrub makes a really unusual choice in a sunny, sheltered place outdoors. Its chief appeal are the showy, rounded whorls of slender, rather curved, tubular, fiery-orange flowers, which measure around 6cm long and are produced in multitudes through late summer and autumn, encircling the stems and providing the plant with a most distinctive appearance. These blooms emerge from clusters of spiny calyxes that are spaced up erect, if rather brittle stems, the plant as a whole reaching around 2m. The stems are square in cross-section, which helps identify *Leonotis leonurus* as a member of the mint family, Lamiaceae. Lower parts of stems become woody and branched and are well-furnished with slender, lance-shaped, strongly aromatic, deep green leaves; softer upper parts sport paired leaves below each whorl of clustered flowers.

Leonotis leonurus is a native of South Africa, where it is commonly named wild dagga and known for various medicinal properties; its aromatic leaves, for example, are sometimes used to make a calming infusion. It is found growing in sun in dry grassland, forest margins and rocky areas, and in the wild is largely pollinated by sunbirds – the nectar-rich flowers matching perfectly in shape the form of the bird's beak. Happily pollinating insects will also visit the blooms in places where sunbirds are not found. Yellow- and even white-flowered plants are often cultivated; crosses made with the usual orange-flowered plants provide individuals with interesting cream or peach-tinted blooms. In its native land, lion's tail is a popular garden plant, tolerating spells of drought well, blooms lasting in beauty for several months and attracting wildlife. In largely frost-free Mediterranean climates, it thrives and is easily grown – plants stand temperatures down to freezing. Below this, *Leonotis* is soon damaged, behaving more like a herbaceous plant, dying virtually to the ground. The roots do not tolerate being frozen for long, so while it may survive mild, dry winters or linger in sheltered sites, plants are not reliable outdoors over winter in most temperate climates. Happily for gardeners, this plant is easy and impressively fast-growing from seed, spring-sown in a warm greenhouse and regularly potted on until the cold weather has passed.

S. Edwards del. Pub: by W. Curtis. St. Geo: Crescent May 1. 1800. F. Sansom sculp

Opposite Lion's tail (*Leonotis leonurus*) from *Curtis's Botanical Magazine*, 1800.

Treat it as either an annual or a tender perennial, kept in a large container and bought under glass for winter. It likes fertile, well-drained soil and so is useful in mixed or herbaceous borders in late summer, filling space after earlier flowering plants such as *Papaver orientale* have faded – either plant young *Leonotis* out directly, or partially sink tub-grown plants in the ground. Young plants raised from seed are particularly slender and upright – they get bushier with age. Try using these amid bold, exotic-looking plants such as *Aeonium*, *Agave*, *Canna* or *Melianthus major* – its slender swaying stems of dazzling flower will give the composition a dynamic touch. A rather different feel can be had from interplanting *Leonotis* amid tall grasses such as *Ampelodesmos mauritanicus*, *Austroderia richardii* or *Celtica gigantea*, perhaps in combination with tall-growing *Erigeron annuus* and *Verbena bonariensis*. You'll need to water young plants well to begin with, but as they establish they soon become tolerant of dry, sunny weather. In temperate regions, lift them before the frost or let them take their chances outdoors; positioned by a sunny, sheltered wall and given a thick, organic mulch they may survive several years. Damaged top growth can be cut back hard once new shoots appear at the base in spring.

OTHER PLANTS TO TRY

- ***Phlomis russeliana***, also from the mint family, bears whorls of flowers, in this case smaller yellow blooms held on shorter stems to around 80cm tall in early summer, above a dense ground-covering carpet of broad, hairy leaves. It is an easy, drought-tolerant perennial with the added virtue of attractive seedheads if left over winter. Survives -20°C.
- ***Phlomoides tuberosa* 'Amazone'** is similar to ***Phlomis russeliana*** with taller, more slender spires of soft, lilac-pink flowers appearing in early summer. It makes a more upright-growing, clump-forming plant to around 1.2m. Hardy to -15°C.

Catmint

Nepeta racemosa

Soft, aromatic grey-green foliage topped by airy sprays of violet-blue flowers in summer make this easy perennial perfect for dry, sunny sites, be they formal or wild-inspired.

Opposite Catmint (*Nepeta racemosa*) from *Curtis's Botanical Magazine*, 1832.

This useful summer-flowering, clump-forming herbaceous perennial is a great choice for dry, sunny, temperate gardens. Its spreading but bushy growth and soft, aromatic grey foliage form superb ground cover, and grown well for several months produces whorls of small, tubular, soft mauve-blue flowers, grouped in loose sprays. The high point of the display usually arrives in early summer when the mounds of growth are covered in shimmering flowers, attracting bees and other pollinators from near and far.

This plant is native to Turkey and parts of the Middle East and comes from stony, arid, sunny places and so it has a tolerance of poor, dry conditions – it reaches around 50cm tall and as much across, although sizes of selections differ. Together with Mediterranean *N. nepitella* it is a parent of vigorous, larger growing *Nepeta x faassenii*, a garden hybrid known since the end of the eighteenth century. All revel in poor, freely-draining soil, especially light, sandy ground, enjoying open sites in full sun and standing considerable periods of summer drought with ease. In winter, stems die back completely to ground level; plants are fully hardy, tolerating temperatures below -20°c as long as soil is not too wet; come spring after old growth has been trimmed away fresh shoots soon emerge, initially forming a neat clump but as stems elongate the habit becomes more spreading.

Nepeta racemosa proves versatile, fitting brilliantly into many styles of garden; planted en-masse it can be used as a soft hedge to line a path or avenue for more formal schemes, while on a sloping site it can tumble down a bank or soften terrace edges. It makes a superb frothy underplanting for many shrubs but particularly roses, hiding unsightly lower stems, and fills annoying spaces right at the front of borders, helping to hide path edges. In gravel gardens it forms soft clouds of blue helping to knit together shrubs and herbaceous components. The plant's soft pastel tones mean it blends with all manner of plants. It's a great choice for a cool border of silvers, whites and blues, alongside *Artemisia*, *Agapanthus*, *Cistus x hybridus*, lavenders, *Salvia yangii* and *Teucrium fruticans* and it contrasts beautifully with pastel *Cistus* 'Silver Pink', pale-yellow flowered *Rosa x odorata* 'Yellow Mutabilis' or *Hylotelephium* Atropurpureum Group 'Purple Emperor'.

Opposite *Nepeta nepetella* from *Deutschlands Flora in Abbildungen nach der Natur mit Beschreibungen* by Jacob Sturm, 1841.

Choose selections such as 'Walkers Low' which stays relatively bushy with violet-tinted flowers all summer. For white blooms choose 'Snowflake', a great plant for a cool-themed planting palette, as is delightful pastel pink 'Amelia'. Blue 'Little Titch' is perhaps the most compact, reaching 20cm tall; perfect for the front of borders. In truth, most selections of *Nepeta racemosa* are compact; most about half the size of popular N. 'Six Hills Giant' which means they fit easily into existing planting and beside a path they edge rather than obstruct. They do not run at the root, unlike *N. sibirica* 'Souvenir d'Andre Chaudron', which while charming can in time spread rather too freely. Seedlings may occasionally arise but gardeners usually increase plants through division of clumps in spring. While easy-to-grow, this perennial is best with some attention through the year to keep it looking good. Once the initial display has finished by mid-summer, cut plants back hard to promote fresh new shoots that flower again before the season ends.

The most common garden pest with this plant is likely to be a furry one; most (but not all) cats find *Nepeta* utterly irresistible – hence the common name. They love to roll amid its stems and nip at shoot tips, the foliage containing a chemical that induces a state of euphoria in them; you may be amused by this kitten-ish behaviour and tolerate the damage, or decide to put them off by placing cut stems of prickly holly or rose amid the clump.

OTHER *NEPETA* TO TRY

- ***Nepeta grandiflora* 'Dusk till Dawn'** makes an appealing garden plant with spires of white flowers, 80cm tall, emerging from contrasting reddish calyces. Hardy to -20°C or less.
- ***N. sibirica* 'Souvenir d'André Chaudron'** makes a superb summer display, the charming, rather large, tubular glowing blue flowers nodding from whorls around upright stems above pointed pale green leaves, to 1m tall. It spreads at the root and stands drought when established. Hardy to -20°C or less.
- ***N.* 'Six Hills Giant'** is perhaps the best-known catmint, and among the most spectacular, forming clouds of silver-grey growth topped by sprays of small soft blue flowers to around 1.2m tall and more than 2m across, so it needs space. Hardy to -20°C or less.

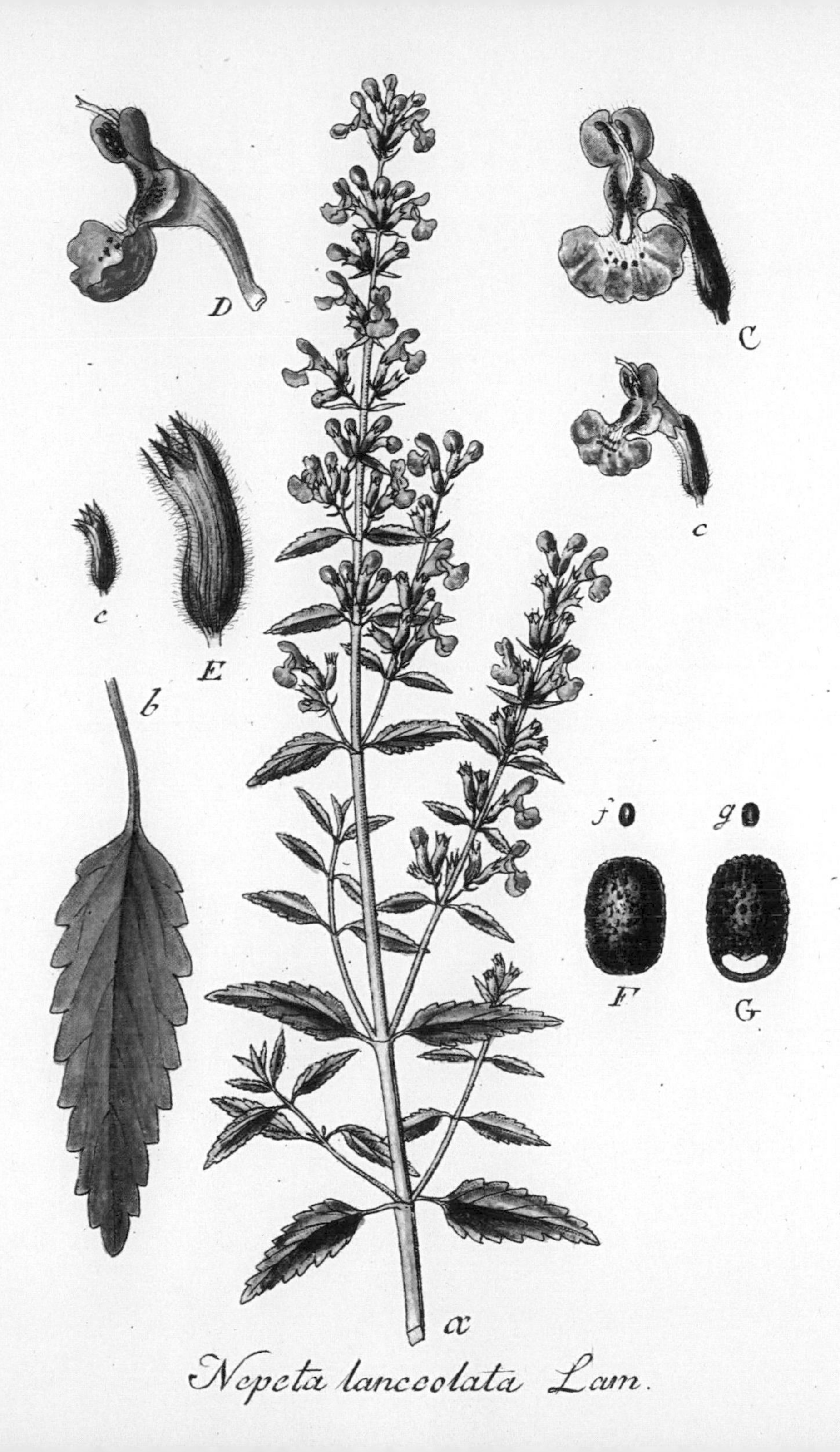

Nepeta lanceolata Lam.

Bowden lily

Nerine bowdenii

This autumn-flowering bulb from South Africa produces impressive heads of vibrant pink flowers just as the garden is beginning to fade, flourishing in hot, dry summer conditions.

The early part of autumn so often feels like a relief in the garden. Days that were baking hot, and by consequence often bone dry, are now behind us; the weather is fresher but still mild; and with luck there is occasional rainfall, meaning that many plants now come back into growth, the garden at times feeling almost spring-like once again. Enhancing this welcome illusion are bulbous plants adapted to flower as days shorten; *Colchicum*, *Sternbergia*, some *Galanthus* and *Crocus*, and queen of all, lovely *Nerine bowdenii*.

Opposite Bowden lily (*Nerine bowdenii*) from *Curtis's Botanical Magazine*, 1807.

There are around 25 species of *Nerine*, all native to South Africa. Some, such as well-known *N. sarniensis* (Guernsey lily), flourish in Mediterranean regions, basking in late-season sun and heat, but these are too tender for temperate climes, their foliage damaged by frost. A better bet is *N. bowdenii*, which is far more tolerant of cold, coming from the Eastern Cape, where frost is common, and standing perhaps -15°C. More importantly, it is also adapted to summer heat and drought. It was introduced to widespread cultivation around the end of the nineteenth century and has been a firm favourite with gardeners ever since – there are now numerous named selections with distinct attributes and some succeed in gardens as far north as central Scotland, given a little shelter and free-draining soil. Plants begin into growth in late winter (some selections start later in spring), producing from the noses of their bulbs, sturdy, glossy-green, strap-shaped foliage which endures through summer until flowering time in autumn. Spear-like flower stems rise to around 40cm (more in some selections), the flower buds protected by a pointed, papery sheath. The buds open into a showy umbel of usually dazzling pink flowers (white and other rosy shades are also available) – at least eight per stem but often 12 or more. Individual flowers reach around 4cm across and have slender, reflexed, rather frilly petals that sparkle in the autumn sun, a marvellous sight, grown well.

In Mediterranean gardens and even sunny and sheltered, temperate sites, *Nerine bowdenii* may be easily grown in the open garden, although in cooler climes plants usually perform best against a sunny wall, where they can soak up reflected summer heat and enjoy some winter protection. Bulbs do best in reasonably fertile soil, but it must be well drained – a raised

NERINE BOWDENI.

Opposite Bowden lily (*Nerine bowdenii*) from *Flora and Sylva* by William Robinson, 1905.

bed is a great place; poor, sandy soil will also do, but it is worth adding garden compost at planting time – generally spring when bare root bulbs are available. Unusually the bulbs are positioned with the tips just above soil level. Once planted, bulbs are best left well alone – they flower best with minimal disturbance and will not need any extra water through summer. Mulching is not recommended because it may cause the bulbs to rot.

Finding planting partners is tricky: these plants don't do well with competition. It is crucial to keep *Nerine* leaves in good condition and in sun throughout summer – shade from other plants reduces flowering and leaves can be damaged by slugs and snails that hide in nearby foliage. Consider trying with *Anemonoides blanda*, *Cyclamen hederifolium* and *C. coum*, which are all largely dormant while *Nerine* foliage develops. Smaller annuals may be worth a go; a scattering of *Myosotis* (forget-me-not) or *Eschscholzia californica* for example. In recent years the number of *Nerine bowdenii* selections available has increased, partly through interest in these plants from the cut flower industry. Some distinct kinds that have proved highly garden-worthy include white 'Perla Blanca'; reliable 'Isabel' with strong pink blooms; profuse, rich pink 'Regina'; and charming 'Stephanie', its pale pink blooms with a hint of peach.

OTHER *NERINE* TO TRY

- ***Nerine undulata*** is less often grown than ***N. bowdenii***, but can make a fine garden plant given a really well-drained place, perhaps in a sunny rock garden. The smaller later flowers and shorter stems make it distinct – the blooms usually pale pink and rather spidery in appearance. The selection '**Seaton**' is a particularly fine, pink-flowered plant, while also rather marvellous is ***N. undulata*** (Flexuosa Group) '**Alba**' with shining, pure white flowers late in the season which stand out amid the turning leaves. Hardy to -10°C.

Oleander

Nerium oleander

Among the most drought-resistant of shrubs, this dazzling yet toxic plant flowers almost year-round in Mediterranean countries; hardier selections are worth trying in cooler climes.

Opposite Oleander (*Nerium oleander*) from *Paxton's Magazine of Botany, and Register of Flowering Plants* by Joseph Paxton, 1837.

Travel along any Puglian motorway on a blazing August day, and you are likely to see banks of a tall flowering evergreen shrub, usually with strong pink, but also white, peach, yellow or red, single or double blooms, carried in impressive clusters on upright or arching stems. Seemingly impervious to the heat and drought, this is oleander, a plant just as synonymous with Mediterranean climates as *Bougainvillea* and prickly pear; indeed, the wild origins of this plant have been lost, as it has been cultivated so widely, for so long. This long-standing popularity is largely due to its remarkable tolerance of drought and extreme heat, yet increasingly this plant is being tried out – and succeeding – in cooler, temperate areas where warm summers and milder winters are now experienced. It's also highly versatile; it forms common municipal planting in hot, dry climates, growing almost anywhere. Occasionally plants are even trained on single stems and used as small street trees.

Out of flower (briefly in hot climates) *Nerium oleander* is a fairly unexciting plant with long, dark green lanceolate leaves, each with a paler midrib. Young plants tend to be rather erect and with few branches. The branched panicles of flowers are carried in erect terminal shoots which appear in spring and gradually develop, opening from pointed buds through summer. Plants can reach 5–6m tall and much the same across, and are multi-stemmed, the outer shoots splaying over to reveal bare lower growth, meaning that older examples can be rather awkward-looking, a characteristic solved easily by occasional pruning. Individual flowers are beautiful: propeller-shaped, usually with five petals and around 4–5cm across – in isolation they look rather like the blooms of periwinkle (*Vinca*) which is a member of the same family, Apocynaceae. Double-flowered selections are common and make an even more impressive show, their branches often arching over through the sheer weight of bloom. These also often have a pleasing, sweet coconut or vanilla-like perfume – a characteristic seldom encountered in single-flowered plants (some white-flowered plants are also perfumed). Plants flower profusely in warmer climates – often more or less year-round; in cooler climates just once in mid to

Opposite Oleander (*Nerium oleander*) from *Traité des arbres et arbustes qui se cultivent en France en pleine terre* by Duhamel du Monceau, 1812.

late summer. The blooms may be followed by long and slender, pointed seed pods, which split open when ripe to reveal fluffy seeds that are distributed by the wind.

Oleander certainly has many attractive virtues, but there are two snags. The first is that the plant is highly poisonous. There are old (probably fictional) tales of people confusing *Nerium* foliage with *Laurus nobilis* (bay) and using it to season food, with disastrous consequences. However, it is also said to be revoltingly bitter to taste, meaning that poisoning is very rare, even in countries where the plant is ubiquitous. It is certainly a plant to be treated with respect – any substantial handling or particularly pruning must always be done using gloves and it should be kept out of the reach of children or pets.

The other problem is hardiness; most selections will tolerate only brief, light frost. These can still be enjoyed outdoors in temperate countries if kept in large containers – they prove easy to care for and need far less water than many other potted plants. In winter they can be moved into a conservatory or greenhouse – plants are best above 5°C and kept on the dry side, rather as you might grow *Citrus* or *Brugmansia* plants. That said, there are cultivars (mostly single-flowered) which are proving far hardier, surviving temperatures of -5°C or slightly less and flowering well in mild locations such as coastal or city gardens, grown in full sun with the shelter of a warm house wall in well-drained soil.

OTHER *NERIUM* TO TRY

It is well worth experimenting with tougher single-flowered oleanders in sheltered parts of temperate areas. For selections suited to near frost-free gardens or for growing in tubs kept under glass in winter, the choice includes '**Alsace**' (single, white blooms); '**Angiolo Pucci**' (single, orange-cream); '**Madame Allen**' (double, light pink); '**Maresciallo Graziani**' (single, salmon pink); '**Oasis**' (double yellow); '**Rubis**' (double, dark red); and '**Roseum Plenum**' (double, red-pink).

NERIUM Oleander. **NERION** Laurier-rose.

Olive

Olea europaea

With shining silvery foliage, this long-cultivated tree has proven in recent years to be a tough and versatile ornamental species, thriving even in cool, temperate gardens.

A few decades ago, gardeners in frost-prone, temperate lands would have given little credence to the idea that olive trees could soon become relatively widespread ornamental plants in many areas. Today they are often planted in places formerly considered far too chilly for these denizens of droughty, sun-blasted Mediterranean lands, proof perhaps of our changing climate. Few trees are more characterful than a mature olive; they can live for more than a thousand years; some of the real old-timers surviving in parts of Italy and Spain are thought to have been planted as saplings during the final years of the Roman empire. After just a few decades they begin to develop a marvellously craggy, gnarled grey trunk, which supports an evergreen canopy of leathery green leaves. These have conspicuous silver undersides, with trees at their shimmering best in spring as new growth begins – an impressive sight under a cloudless azure sky. In summer, small cream sprays of little flowers develop and these go on to produce olive fruit that will begin to form even in cooler climates, although they need consistent heat to mature and soaking in brine after picking to make them edible.

Choose a really well-drained site in full sun – a site at the top of a slope or perhaps in a terraced garden near a retaining wall. Young plants also do remarkably well in pots for a while. You will need to keep them watered as they establish, but an olive won't need extra irrigation once settled. Old mature trees grubbed out from redundant groves and containerized are frequently offered for sale at huge prices, but in truth young, healthy saplings soon make handsome trees and establish quickly without the financial risk. Plants should survive around -10°C once established – some defoliation may occur in the coldest conditions.

Olives make the perfect specimen tree for a gravel garden, sunny terrace or patio – anywhere a Mediterranean feel is desired – and they fit into traditional and contemporary gardens equally well. They look great with other silvery and typically drought-tolerant plants – contrast their rounded crowns with the dark columns of cypress (*Cupressus sempervirens*), grow beside *Cistus*, lavenders, *Phlomis*, rosemary and salvias and allow

Opposite Olive (*Olea europaea*) from *Plantae Medicinales* by Nees von Esenbeck, 1828.

A
B
a
c
d
e

Opposite Olive (*Olea europaea*) from *Flora Graeca* by John Sibthrop and James Edward Smith, 1806.

climbing *Clematis cirrhosa* or *Solanum laxum* to scramble through their branches. Perennials such as *Cyclamen hederifolium*, *Nassella tenuissima* and even bold *Melianthus major* will usually flourish below, especially if you occasionally thin the canopy. In fact, olives seem remarkably adaptable and stand plenty of pruning. For commercial olive production, branches are well-thinned to allow for maximum fruit ripening – in parts of Italy it is said that a swallow should be able to glide with ease through a well-pruned olive tree's crown. Increasingly, though, trees are sold pre-shaped for aesthetic purposes – young plants are trained simply into standards, while older plants get a more intensive makeover; some of the best end up resembling giant bonsai, others appear closer to vegetable poodles, rather an indignity for such a noble tree. In truth, the natural form of the tree suits most settings perfectly, but it is good to know that should a tree outgrow its place it can be kept within bounds without worry.

Sadly, olives are under threat. The devastating plant disease *Xylella*, which quickly kills even old, mature trees and can affect a wide range of other plants, is laying waste to once beautiful olive groves, especially in the south of Italy, and heading north with alarming speed. Many trees sold in garden centres and nurseries around Europe and further afield will have been originally raised in Italy, and there is concern that despite phytosanitary measures the international trade in plants could cause it to spread further.

OLIVE ALTERNATIVES

- ***Elaeagnus* 'Quicksilver'** forms an arching shrub or small tree and bears shimmering silver foliage. It is deciduous, unlike a true olive and has scented yellow flowers. Choose a sunny, open site; plants tolerate dry spells when established.
- ***Tamarix tetrandra*** forms an open shrub or tree, similar in shape to an olive and standing heat, full sun and lengthy drought superbly. The plume-like foliage is distinct, as are the panicles of pink summer flowers.
- ***Phillyrea latifolia*** resembles an olive in form, with dark evergreen foliage held on arching branches in a dense crown, as well as white spring flowers and small blue fruits. Sun-loving and drought-tolerant once established, it is slow growing but stands exposed cold sites.

Prickly pear

Opuntia ficus-indica

A heat- and drought-proof cactus with segmented stems and crops of colourful, succulent seed-filled fruits. While tender, it makes a great talking point for a patio pot in cool climes.

Opposite Prickly pear (*Opuntia ficus-indica*) from *The Cactaceae* by Nathaniel Lord Britton and Joseph Nelson Rose, 1919.

This cactus will likely be familiar to anyone who lives in, or who has visited, almost any arid but frost-free land. Probably native to Mexico, it has been cultivated in South America for centuries and, since its introduction to Europe in the sixteenth century, has become the most widely grown of all cacti, usually for its sweet, fleshy, seed-filled fruits. It is also used as fodder for livestock, and to stabilize and improve ground in places that prove too dry and poor for other plants, reducing erosion and, in time, allowing other plants to grow. In parts of some countries such as Australia, the plant is highly invasive.

There are around 300 species of *Opuntia*, although they freely hybridize, so the true figure is hard to judge. They are easy to identify, thanks to their highly distinctive, unusual segmented stems. These are composed of flat, rounded leaf-like pads (cladodes) which have a thick, waxy cuticle and are usually armed with fixed, thorn-like spines and smaller prickles that easily detach and serve as an irritant should they become embedded in skin. *Opuntia ficus-indica* can reach 5m; the plants may be upright and tree-like with a distinct trunk and crown, but in time may collapse under their own weight. The 60cm-long, blue-green pads, as well as the sections of stem that drop off, will take root below, forming a sprawling hedge-like mound of impenetrable growth; these often serve as field boundaries or flank roadsides. All (or almost all) cacti hail from the New World; numerous *Opuntia* are native to the United States. They are brilliantly adapted to heat and drought, while a few also come from high altitude or northern areas with considerable winter cold, including snow and frost. Although thriving in extreme heat, *Opuntia ficus-indica* is not a species tolerant of frost; plants are best above 5°C.

The beautiful flowers of prickly pear are yellow or even orange-tinted, around 10cm across, produced on edges of upper pads in spring and summer; these develop into fig-shaped fruit that are yellow, orange, pink or purple when ripe. Sometimes called 'tunas', these are tempting in the searing heat, but the uninitiated soon learn that the skins are also covered in fine spines, which makes eating them a fiddly, painful experience

Opposite Prickly pear (*Opuntia ficus-indica*) from *Phytanthoza Iconographia* by Johann Wilhelm Weinmann, 1742.

seldom repeated without precaution. Fruits are also made into tasty drinks and jams. In southern Italy, pads with several fruits are detached and hung on a cool wall – this preserves fruit for months. In Sicily, extra-large fruit known as *bastardoni* are produced by removing spring flowers, promoting later blooms followed by fruit that develops in autumn, when wetter weather allows them to swell impressively. Spine-free selections are also grown, which makes eating them easier. In some places, young pads are even eaten as a vegetable, cut into strips and cooked.

Prickly pears are common garden plants in many hot, dry countries, but often found at the perimeters of properties, growing over dry walls, perhaps alongside *Agave americana*.

In temperate climates, *Opuntia ficus-indica* can grow well and even produce occasional fruit kept in a greenhouse, while small plants also make highly attractive, rather architectural specimens for standing on a sunny terrace or patio through summer, contrasting with other succulents and drought-tolerant plants – they look superb planted in a terracotta urn or pot beside an olive tree, providing the ultimate Mediterranean accent. Over winter, keep them virtually bone dry in a heated greenhouse, conservatory or on a sunny windowsill – if they eventually get too large, simply detach a pad and root it in a pot of gritty compost. Some other species are more cold-tolerant, even standing winters outdoors with some shelter.

OTHER *OPUNTIA* TO TRY

- ***Opuntia englemanii*** var. ***cuija*** is a species with rather rounded blue-green pads and low sprawling stems to around 1.5m, which proves impressively cold-tolerant if grown in sharply draining soil, perhaps on a slope and with protection from excess rainfall. Hardy to -8°C.
- ***O. erinacea*** (old man prickly pear) is distinctive for its long, white, rather fine, weak spines on stems to around 1m high. It is quite a hardy species if kept bone dry in winter. Hardy to -8°C, possibly lower.
- ***O. humifusa*** is one of the best-known hardy species, to around 1m high, standing even snow and winter wet – ideal for a sunny, well-drained rock garden, where it may even produce yellow flowers and edible fruit. Hardy to -10°C, possibly lower.

c
c
a
b
a. Opuntia seu Ficus In=
dica maxima, Raquette,
grosse Indianische Feigen.
minor seu Ficus Indica ,
nische Feigen. c. Opuntia minor
epineuse petit, kleine stachlichte
b. Opuntia,
Cardasse, kleine India,
spinosissima, Figuier
Indianische Feigen.
K.

Rose-scented geranium

Pelargonium capitatum

With powerfully aromatic foliage and heads of pretty pink flowers, this tender perennial is a great quick-growing choice as a seasonal plant and stands hot, dry conditions well.

Loved for its deliciously aromatic foliage and heads of small but showy flowers, this South African native has long been enjoyed as an ornamental plant in both temperate and frost-free climates, but this is a highly useful plant too; rubbing its foliage on cracked skin is said to soothe and soften, while people from the Cape use it to make an infusion to calm digestive problems. It is also now grown commercially as a source of essential oils for use in aromatherapy or for making topical skin creams. In the UK, the Victorians often cultivated it under glass, enjoying its rose flavour in cakes and jellies, and adding sprigs to posies and potpourri for scenting rooms. You can even make cordials from it or add the leaves to cocktails.

Rose-scented geranium is a compact, shrubby plant, reaching around 1m or so tall and a little more across. Its fresh green, rather crinkled, lobed foliage and stems are hairy, and when touched give off a sweet scent that can be highly reminiscent of the perfume of rose flowers. In the wild it grows in open, coastal areas in sandy, well-drained soil – thriving in sand dunes and in fynbos (species-rich shrubland vegetation unique to the Western and Eastern capes) with proteas, ericas and restios.

Pelargonium capitatum is one of many scented-leaf pelargonium species and a parent of numerous selections; it is also among the simplest of all to grow. The attractive, usually pink or mauve-purple blooms have petals with interesting darker markings; each flower is 2–3cm wide and clustered in neat, rounded heads of up to 20, held on short stems clear of the foliage. Unlike the leaves, they have no scent but may be followed by small, spiked seedheads. The plant was introduced to cultivation in 1690 and has since proved a highly popular and adaptable species, standing spells of drought and neglect well; in parts of Australia, where conditions in the wild are similar to those found in its South African home, it is now classed as an invasive weed. In many Mediterranean gardens it is often seen cascading from pots and window boxes on balconies or terraces, or forming softening yet dense ground cover alongside plantings of succulents and cacti, or below shrubs such as oleander and

Opposite Rose-scented geranium (*Pelargonium capitatum*) from *Traité des arbres et arbustes qui se cultivent en France en pleine terre* by Duhamel du Monceau, 1819.

Opposite Rose-scented geranium (*Pelargonium capitatum*) from *The Romance of Nature* by Mrs Charles Meredith, 1836.

Hibiscus, perhaps with other drought-tolerant plants such as lantana, lavender and prostrate rosemary. It stands pruning, which helps to keep plants bushy and tidy.

In temperate climes, rose-scented geranium is widely used as a bedding plant or in seasonal container displays. It also makes an excellent and easy to keep house or conservatory plant flowering well indoors for much of the year if kept in a sunny position and regularly deadheaded. Like most other pelargoniums it is a tender plant, surviving only very light frosts which quickly defoliate plants – they will grow new shoots from the base if roots do not freeze, but generally they are best kept above 5°C. Happily for gardeners, these plants are simply propagated from stem cuttings, most easily taken in spring from old plants that have overwintered. Once rooted they grow quickly and make a handsome container plant in free-draining compost, perhaps used as a summertime centrepiece for a herb garden, or perhaps positioned with other scented leaf pelargoniums close to a path or doorway where the fragrance can be easily enjoyed. In containers they generally need far less watering than many other traditional bedding plants, particularly other old favourites such as petunias and begonias – unlike these, they will not suffer if the odd watering is missed, so they prove well-suited to hotter, drier summers and for gardeners who may often be away for a few days and unable to irrigate.

OTHER SCENTED LEAF *PELARGONIUM* TO TRY

- ***Pelargonium* 'Attar of Roses'** is one of the best-known selections of ***P. capitatum***, with strongly rose-scented foliage and pink flowers – it's a perfect choice if you are looking for a rose-scented pelargonium for culinary use. Keep above 5°C.
- ***P.* 'Prince of Orange'** has orange-scented foliage and is another for adding to cocktails, using to make cordials or even flavouring ice cream. Pale mauve pink flowers. Keep above 5°C.
- ***P. tomentosum*** (peppermint geranium) is a handsome plant with lush, broad, almost velvet-like, silvery-green leaves 10cm or more across, which are powerfully scented of peppermint. Sprays of small white flowers. Keep above 5°C.

Canary Island date palm

Phoenix canariensis

This magnificent palm is ideal for landscape use or large gardens in frost-free climates and also in colder areas as a young potted plant for a sunny terrace, overwintered in a greenhouse.

Opposite Canary Island date palm (*Phoenix canariensis*) from *Revue horticole*, 1888.

Of all plants, surely it is palms that most instantly and effectively provide a garden with an exotic, if not tropical, feel – and this noble species proves to be one of the most versatile, surviving in a range of situations in climates from tropical to temperate. Native to the Canary Islands, it is closely related to *Phoenix dactylifera* (date palm), but *P. canariensis* differs in its brighter green, more luxuriant foliage and stout trunk, although hybrids between these species are also common. Mature Canary Island date palms are a handsome sight indeed, their many hundreds of arching, feathery fronds to 5m long forming a dome-shaped crown 10m or more across, atop a brown trunk that may almost reach 1m across on the largest examples. In the wild, the trunk is often host to a range of epiphytic Canary Island plants, including yellow-flowered *Sonchus acaulis*, succulent *Aeonium* and various ferns.

Close up, that soft-looking foliage is far less cuddly, the central midrib of each leaf armed with savage needle-like spines. In spring, plants may bear masses of tiny flowers held on an orange-yellow branched structure held amid the leaves and from these, oval fruit develop, containing a single seed. These seeds germinate readily and in some areas, such as the southern United States, the plant has become naturalized.

This is a spectacular ornamental tree and probably the most widely cultivated palm for garden and landscaping use, making an impactful sight in parks and forming majestic avenues in largely frost-free climates around the world. In these locations it will make a really big, lofty plant when mature, eventually to around 40m tall with a bulky trunk. Even when far younger, it needs plenty of space in which to spread out: it will be years until the trunk is tall enough to walk below, and those spines amid the leaves are best avoided, so don't plant it too close to buildings, drives or pathways. It enjoys full sun or part shade, well-drained soil and thrives in coastal environments, tolerating salt spray well. Until established it will need some watering, but afterwards it is resistant to seasonal drought. It also stands short periods of waterlogging, a useful characteristic in urban situations that suffer occasional flash flooding. Where once these fabulous palms were commonplace, mature examples of

Opposite Date palm (*Phoenix dactylifera*) from *Traité des arbres et arbustes qui se cultivent en France en pleine terre* by Duhamel du Monceau, 1809.

this plant are now becoming a rare sight in many Mediterranean areas, thanks to the ravages of Asian palm weevil, a large red beetle. The female beetle lays eggs on new foliage in the crown of the palm, then the larvae burrow down into the heart of the palm, eating it from the inside out and usually eventually killing even mature trees.

In temperate climates planting *Phoenix canariensis* is, in honesty, a gamble, even in mild areas. It is interesting that even in the extreme south-west of the UK (except the Isles of Scilly) mature specimens with trunks (i.e. palms more than around five years old) are rare – the occasional cold winter every decade or so sees them off. Even mature plants are readily killed by temperatures below -10°C, and juveniles seldom survive -3°C. That said, this palm is so handsome in all stages of growth and so easily obtained that many gardeners in frost-prone areas are often content with youngsters to grace gardens for just a few years. Even plants 30cm high look great in containers of well-drained compost, placed on a hot, sunny terrace, used almost like bedding plants, only drought-tolerant. Offered winter protection in a frost-free glasshouse, they will keep for several years; you can move them outdoors each summer until they get too large to shift. In really mild areas, they can make impressive specimens planted out in a warm well-drained spot until the next inevitable freezing winter happens – although with climate change, these are perhaps becoming less frequent.

OTHER *PHOENIX* TO TRY

- ***Phoenix dactylifera*** (date palm) is similar to ***P. canariensis*** but rather more slender, with blue-green leaves. Grown since ancient times for its crops of sweet fruit, it can also be a clump-forming palm, plants producing side shoots that develop into trunks. Young plants are also easily raised from the stones of fruit. Keep frost-free.
- ***P. roebelenii*** (miniature date palm) may be the species to choose if you are looking for a small, easy-to-manage palm that will stand outdoors in summer but spend frosty winters in a conservatory or as a houseplant. Reaching 2–4m tall with a slender trunk and leaves 1m long, it is perfect in a pot. It likes more moisture than ***P. canariensis***, and stands part shade. Keep frost-free.

PHŒNIX Dactylifera. PALMIER-DATTIER.

Stone pine

Pinus pinea

Wonderful, characterful broad-topped pine planted widely in Mediterranean countries for its handsome appearance, shade-casting qualities and crops of delicious, edible seeds.

Opposite Stone pine cone (*Pinus pinea*) from *A Description of the Genus Pinus* by Aylmer Bourke Lambert, 1832.

This glorious pine is a Mediterranean classic – one of the most instantly recognizable of all drought-tolerant trees and a conifer cultivated in the region for thousands of years, initially due to its edible seeds (pine nuts), but also for its distinctive architectural form. This is the pine many will have admired for its wide-spreading, flat-topped crown, a characteristic that gives the tree its alternative common name, umbrella pine, and which is often useful in providing respite from the searing Mediterranean sun. Ancient Romans certainly appreciated it – in the capital of their empire they planted it to form avenues, shading important roads. As a landscape tree it forms a perfect visual counterpoint with the region's other best-known conifer, *Cupressus sempervirens* (pencil cypress); stone pine's broad, soft outline contrasts perfectly with the slender column-like cypress. *Pinus pinea* is originally a Mediterranean native, found in scrub and woodland in southern France, Spain, Portugal, Italy and Croatia, eastwards to Turkey and Syria and south to Morocco. It is also naturalized in South Africa and Australia, and prized as an ornamental elsewhere, such as in California. It is hardy enough to succeed in temperate parts too, surviving lows of around -10°C.

Umbrella pine reaches around 20m tall, occasionally a little more. Taller, straighter examples are usually those that have been drawn up through competition with nearby trees – in the open this pine tends to be on the stout, stocky side. The dense canopy will eventually be almost as broad as the tree is tall, carried atop a stout, often sinuous, single, reddish-brown trunk. By contrast, young plants are rounded, usually of shrubby, multi-stemmed appearance until a leader is eventually formed. Plants reach around 3m in five years when growing happily. Lower branches are shed naturally, but the tree's appearance is improved if these are removed gradually as it grows. Eventually, middle age spread kicks in and the tree begins to branch out, forming the characteristic spreading shape. The foliage also alters – younger trees have greyish blue-green solitary needles, each with a sharp point; after 10 or so years, bright green, paired, 15cm-long needles take over. Fresh shoots in late spring are attractive and like other pines appear as erect, initially yellowish

a
b
c
d
e
f
g
h
i
k
l
m
n
o
p
Le Pin.
Pinus Pinea. Linn. Sp. Pl.
Ital. Pino. Angl. Mountain Pine. Allem. Berg-Zirbel Baum.

Opposite Stone pine (*Pinus pinea*) from *La Botanique mise à la portée de tout le monde* by François Regnault, 1770–80.

candles of growth. The glossy brown seed cones, 15cm long, first appear around the time of adult foliage production. They take three years to ripen, when the rounded scales split apart to reveal oil-rich seeds within. These have a delicious nutty flavour and have long been prized – they are an important ingredient in pesto sauce.

In Mediterranean areas this tree makes a useful windbreak by the coast and may be used to help stabilize sand dunes – it proves both wind- and salt-resistant and flourishes in poor soils, standing heat and long, dry periods. In larger gardens it often forms shelter belts, helping to provide a protected microclimate where more delicate plants may thrive – you also often see *Pinus pinea* planted to provide shade for nearby houses.

In more temperate regions stone pine is not a common tree, despite its relative toughness and reasonable frost-tolerance, yet it makes a handsome garden specimen given space. It provides an immediate feel of warmer climes, a perfect backdrop for arid-style plantings or a specimen tree in a large gravel garden, perhaps with pencil cypress, *Hippophae rhamnoides* (sea buckthorn), *Genista aetnensis* (Mount Etna broom) and eucalyptus. Choose a sunny, open space and well-drained soil, and plant a young tree if you can – semi-mature specimens tend to be top-heavy and prone to wind rock unless substantially staked.

OTHER *PINUS* TO TRY

- ***Pinus aristata*** (bristlecone pine): potentially among the longest-lived of all plants, this slow-growing small pine to around 6m is native to arid parts of North America such as Colorado, the tightly clustered dark green needles spotted characteristically with white resin. Hardy to -20°C or more.
- ***P. mugo*** (mountain pine) is an easily grown, compact conifer to 3–6m, ideal in sun or part shade and thriving in exposed, windy sites. There are numerous named selections, some smaller than others, many with gold-tinted foliage. Hardy to -20°C or more.
- ***P. nigra*** (Austrian pine) is great as a coastal windbreak, a tough pine that grows tall, potentially to around 50m, forming a conical tree revelling in sun with good drought tolerance, yet standing severe frost and snow. Hardy to -20°C or more.

Californian tree poppy

Romneya coulteri

With dramatic, white poppy-like blooms held atop tall, upright stems, this spectacular hardy perennial is a true garden monarch that loves hot, dry locations in free-draining soil, and throughout its flowering season in summer it will be a highlight of any garden.

For many gardeners, dealing with hot, dry sites can be at times difficult, though there are certainly positive aspects, one being that these conditions open up the possibility of growing plants that would never succeed in shaded gardens. High on the list of the most spectacular perennials for a dry garden is Californian tree poppy or Matilija poppy (*Romneya coulteri*), a plant which has been described as the 'queen of Californian flowers'. Native to Californian chapparal, a plant community enduring conditions similar to the Mediterranean maquis, it is a suckering shrub strictly speaking, and notably woody at the base, but in cooler, temperate climates established plants are usually treated by gardeners as a herbaceous perennial, pruned down to the ground at the end of each season. In the wild, it grows in a restricted area of rocky slopes and valleys in the Santa Ana mountains, in near-desert habitat. The rather smaller-flowered *R coulteri.* subsp. *trichocalyx*, a plant less often encountered in cultivation, has a wider range that extends further south into Mexico. In the wild, *Romneya* are known for appearing in recently burned areas of chapparal. The native North American Cahuilla people make a drink from the plant's sap.

Romneya coulteri was introduced to the UK around 1875 yet has never been common in cultivation. In gardens it can be a truly mesmerizing sight in flower, with masses of poppies produced for several weeks through summer, often continuing into autumn. The huge upward-facing blooms, which are sweetly scented and open to around 15cm across, are composed of usually six pure white, rather crumpled petals that look almost as if made of tissue paper, surrounding a luxuriant boss of golden stamens – unsurprisingly, another of its common names is fried egg plant. The flowers open in mid to late summer atop slender growth that may need additional support in exposed gardens, reaching around 2.5 m tall. The stems will branch low down occasionally but do so more often at their tips, each branchlet terminated by a rounded, yellowish flower bud. The foliage is waxy and silver-grey, leaves divided into lance-shaped lobes – in a mild winter the leaves may be retained. The plant forms a large clump and spreads freely at the root, runners

Opposite *Romneya coulteri.* subsp. *trichocalyx* from *Curtis's Botanical Magazine*, 1905.

Opposite Californian tree poppy (*Romneya coulteri*) from *Gartenflora*, 1891.

travelling easily and quickly in light, sandy soil, even spreading under and dislodging paving. It likes an open site, although in colder areas it does well in a border backed by a south-facing wall, where it can bask in reflected heat.

It can be a tricky plant to get established, one reason for it not being grown more widely; if the conditions are not to its liking, the plant may fail, but in the right place it makes a spectacular success. In a dry garden the key with this plant is to get it well established before subjecting it to drought, which it can shrug off easily once the roots have spread out well. Through winter the tips of stems tend to die back, and need cutting back to keep the plant in good order – at least by half but, better still, to soil level to promote new shoots. This obviously leaves a large gap in a border, which can be filled by interplanting smaller, early spring-flowering bulbs, such as *Crocus* and *Scilla*; early interest annuals such as *Myosotis* and *Cerinthe* make other great options. Larger planting partners through the season are easy to find; its upright form and silvery sheen means it sits well behind the rounded evergreen domes of *Cistus* and *Phlomis fruticosa* or beside taller, leafy grasses such as *Miscanthus*, but it is important to give the tree poppy plenty of space and air. It looks good in a gravel garden, where its spreading habit suits the relaxed natural planting style, but it won't tolerate growing in a container for long.

These tree poppies attract pollinators freely, especially bees. Propagation is best achieved from root cuttings taken in winter – detaching a rooted side shoot from an established clump almost always fails.

ANOTHER *ROMNEYA* TO TRY

- ***Romneya* 'White Cloud'** is sometimes offered and thought to be a hybrid between ***R. coulteri*** and ***R. coulteri*** subsp. ***trichocalyx***; it produces even larger flowers, carried more profusely over a longer season.

Beach rose

Rosa rugosa

This tough and versatile, disease-free rose grows in a wide range of challenging conditions, including dry, poor soil, yet flowers all summer and provides showy crops of hips.

Opposite Beach rose (*Rosa rugosa*) from *Edwards's Botanical Register*, 1819.

Roses are not plants often associated with tolerating heat and drought, or even for growing in less than ideal conditions. However, once established, some make really tough garden plants and are found thriving in surprisingly hot climates – even in the tropics and arid parts of the Middle East. One of the most resilient and versatile of all is, without doubt, marvellous *Rosa rugosa*, a species originally native to eastern Asia, including parts of China and Siberia, where it grows in coastal regions, thriving even in sand dunes and on beaches with full maritime exposure, and tolerating long, hot, dry summers as well as freezing winters with ease, hardy to temperatures of -20°C or lower. It has been introduced in countries around the world and is naturalized in many parts of Europe (including the UK) and the Americas, and is even considered invasive in some places, quickly outcompeting native plants and spreading through suckers and seed.

This rose is a deciduous shrub to around 1m tall, sometimes rather more, and has a strongly suckering nature, the plant spreading and forming dense thickets of growth in time. The rather cane-like, erect stems are densely armed with rather short spines, and through spring and summer bear rich-green leaves composed of usually five to seven leaflets. In autumn this foliage turns to attractive bright yellow and orange tones before falling. Unlike many other wild roses, beach rose has an impressively long flowering season, the pollinator-attracting blooms produced generously throughout the growing season, beginning in spring with the new growth but often continuing well into autumn. The beautiful flowers are large, single and of rather delicate appearance, its petals creased rather like tissue paper, and with a delicious sweet fragrance. Blooms are usually magenta-pink, with a mass of golden stamens in the centre, but there is also a very fine white-flowered selection, *Rosa rugosa* 'Alba', not to mention many excellent hybrids closely involving this species with blooms in pink, white and mauve, some double-flowered.

The virtues of *Rosa rugosa* do not end here, for it also produces remarkable displays of large, succulent, glossy orange hips that are edible (raw or used in jams and preserves) and look almost like small tomatoes. They may even appear on the plant

Opposite Beach rose (*Rosa rugosa*) from *L'Illustration horticole*, 1871.

alongside more flowers, a great late-season attribute and one which means that deadheading should be avoided because this stops the hips forming.

Roses are often accused of being susceptible to disease, but this species is also almost immune to the usual rose problems such as blackspot and rust.

In gardens with poor, dry soil – whether sandy or heavy clay – this rose will thrive, and when established stands long spells of drought, growing in full sun, but also tolerating partial shade for some of the day. It is often used by landscapers, mass-planted in municipal areas because it is one of few plants almost guaranteed to survive with minimal care and upkeep, although it looks and performs far better with some attention. Grow it to form a clump in a dry gravel garden, perhaps contrasting its rounded form with upright *Verbena bonariensis, Eryngium* or the architectural outline of *Yucca gloriosa*. Carpeting *Petrosedum rupestre* or low mounds of *Armeria maritima* (thrift) would look good positioned in front. Alternatively beach rose also makes a fine, low-maintenance flowering and fruiting hedge, its suckering habit forming a dense, thorny boundary. Pruning could not be simpler – just thin out the oldest individual stems and reduce the tallest stems by around a third to shape the plant in winter. When planting, young plants will need watering in the first year to establish.

OTHER *ROSA* TO TRY

- ***Rosa* 'Fru Dagmar Hastrup'** is a tough *rugosa* hybrid with large, single light-pink flowers through summer, followed by showy red hips. 1.5m. Hardy to -20°C or lower.
- ***R. glauca*** (red-leaved rose) is worth trying in dry gardens because it stands periods of drought well and succeeds in part shade. It has wonderful red-tinged, silver-green foliage and delicate pink single flowers. 2m or more. Hardy to -20°C or lower.
- ***R.* 'Hansa'** bears well scented double velvety-magenta flowers, followed by fine hips. It's another tough *rugosa* hybrid. 1.5 m. Hardy to -20°C or lower.
- ***R. rugosa* 'Alba'** is perhaps the pick of the bunch with large, pure white blooms all summer and superb displays of showy orange-red hips. 1.5 m. Hardy to -20°C or lower.

Common rosemary

Salvia rosmarinus

Popular and useful in the garden and kitchen alike, this well-known evergreen shrub instantly brings a touch of the Mediterranean to temperate sites with its aromatic foliage.

Of all herbs grown in gardens, rosemary must be among the most versatile – in the kitchen, where its sweetly aromatic foliage combines well with a broad range of flavours, and for its many useful ornamental qualities, which are often rather overlooked. It is another tough, drought-tolerant Mediterranean native, growing wild in open, sunny, well-drained, even sandy sites on hillsides and scrubland, and is quite synonymous with the region, having also been cultivated there for thousands of years, including by ancient Egyptians, Greeks and Romans. Later, in more northern parts of Europe it was grown in monastery herb gardens. It has long been regarded as a symbol of remembrance and sprigs of the plant are still included in funeral wreaths today. The plant is also used to make herbal teas, as well as to flavour meats and vegetables, while oils extracted from its foliage are used in perfumes, soaps and shampoos. With its many uses, rosemary remains a popular garden plant, but in recent years, its naming has undergone a change: the plant is now considered a species of *Salvia*.

Rosemary is typically a dense but rather sprawling low-growing shrub, branches spreading out before becoming more upright, its distinctively fragrant, small, narrowly lanceolate leaves grey-green above but silvery-white below, and held on silvery stems. In gardens, various selections with distinct habits are grown – some with strongly upright growing branches, others with usefully ground-covering or even cascading growth suitable for tumbling over a wall. Plants usually produce small blue flowers – or occasionally white or pink – often in some profusion during spring, or earlier in a mild season, attracting early pollinating insects. These features make rosemary useful for adding early interest to gardens and help explain why this plant has far broader appeal than as a simple component of a traditional herb garden. Plants do vary somewhat in their hardiness; most shrubby kinds will survive -10°C once established, if growing in well-drained soil, but the low-growing kinds tend to be far more delicate, tolerating no more than -5°C. In near-frost free areas, however, trailing selections can form spectacular silvery curtains of growth, the stems cascading for

Opposite Common rosemary (*Salvia rosmarinus*) from *Flora Graeca* by John Sibthrop and James Edward Smith, 1806.

Opposite Common rosemary (*Salvia rosmarinus*) from *Köhler's Medizinal-Pflanzen* by Franz Eugen Köhler, 1887.

several metres. Rosemary makes a highly useful plant for coastal areas because it tolerates maritime exposure well.

Upright growing selections can be easily clipped into low hedges and even simple topiary shapes, such as balls, domes and pyramids, making the plant a good alternative to box hedges in warm, sunny sites. The key is to cut into younger growth only: rather like lavender, plants do not regenerate well from old woody stems. Do this job in spring once the frost risk has passed. Use shaped plants in formal settings such as a small parterre or herb garden, or even simply position a pair of trimmed plants in containers either side of a doorway. Popular 'Miss Jessopp's Upright' develops erect, feathery plume-like growth, which contrasts well with many other more rounded shrubs, such as *Cistus* and *Phlomis*, and looks highly effective in relaxed gravel gardens or areas of Mediterranean-style planting; low-growing types also do well carpeting gravel or softening areas of rocks and paving on a sunny terrace. In particularly cold regions, plants are best positioned by a sunny, sheltering wall while in gardens that have heavy clay soil, these plants may need to be grown in a container or raised bed.

OTHER ROSEMARY TO TRY

- ***Salvia rosmarinus* 'Foxtail'** is a low-growing plant to around 50cm tall, a good choice for tumbling over a wall or the edge of a raised bed. Hardy to -10°C.
- ***S. rosmarinus* 'Majorca Pink'** has an arching habit once mature, reaching around 1m tall and is worth growing for its pretty pink-tinted flowers. Hardy to -10°C.
- ***S. rosmarinus* 'Miss Jessopp's Upright'** is a selection with erect, plume-like stems to around 1.2m tall, bearing in early spring masses of soft blue flowers. Hardy to -10°C.
- ***S. rosmarinus* 'Roman Beauty'** is a shrubby selection to around 1m tall but with a slightly sprawling, arching habit and rich blue flowers. Hardy to -10°C.

Labiatae.
Rosmarinus officinalis L.

Winter daffodil

Sternbergia lutea

Producing wonderful golden-yellow chalice-shaped flowers in autumn, this sun-loving bulb makes an unusual choice and is perfectly adapted to stand summer heat and drought.

Opposite Winter daffodil (*Sternbergia lutea*) from *Floral Illustrations of the Seasons* by Margaret Roscoe, 1829.

Many plants survive adverse seasonal conditions by becoming dormant; perhaps best-known are species that have evolved storage organs such as tubers, rhizomes, corms and bulbs to sustain the plant during the months it retreats below soil level, well away from scorching summer sun or winter frost. Some of the best-known of these bulbous plants bloom in spring, but a few lovely species produce their flowers in autumn; among these is gorgeous *Sternbergia lutea*, a plant closely related to daffodils and agapanthus. This lovely bulb is known for its arresting golden-yellow, chalice-shaped flowers which comprise six petal-like tepals, opening to around 5cm across and held on a slender green stem 10cm above ground level. The bulb starts into growth with the arrival of cooler, wetter autumnal weather, flower buds appearing hot on the heels of strappy, glossy-green foliage that first signals winter daffodil's autumn awakening.

Found wild across the Mediterranean from Spain through Italy and Greece and further east into Asia, *Sternbergia lutea* is at home growing in full sun in stony, well-drained ground; it is often seen on dry, rocky or grassy slopes, usually on limestone soils. It is thought by some historians that plants of *Sternbergia lutea* may well be the original 'lilies of the field' referred to in the Bible. There are around eight species of *Sternbergia* and most have yellow autumn flowers; another is *S. clusiana*, a handsome larger-flowered species, while a couple flower in spring, including *S. candida* with scented and – unusually – white flowers. Both are rather tender, best grown in the protection of an alpine house in temperate climates. *Sternbergia lutea* is, however, an altogether better bet as a garden plant, hardy to around -10°C if grown in well-drained soil. As a result, this bulb has been cultivated for hundreds of years and it has now become naturalized in many parts of Europe – in southern France, it flowers at much the same time as the grape harvest and its golden trumpets can often be seen spangling the ground below olive trees throughout the region.

Before planting *Sternbergia*, it is worth spending time and effort to find the best position. Bulbs must have a free-draining site, so a place in a sunny rock garden or sheltered,

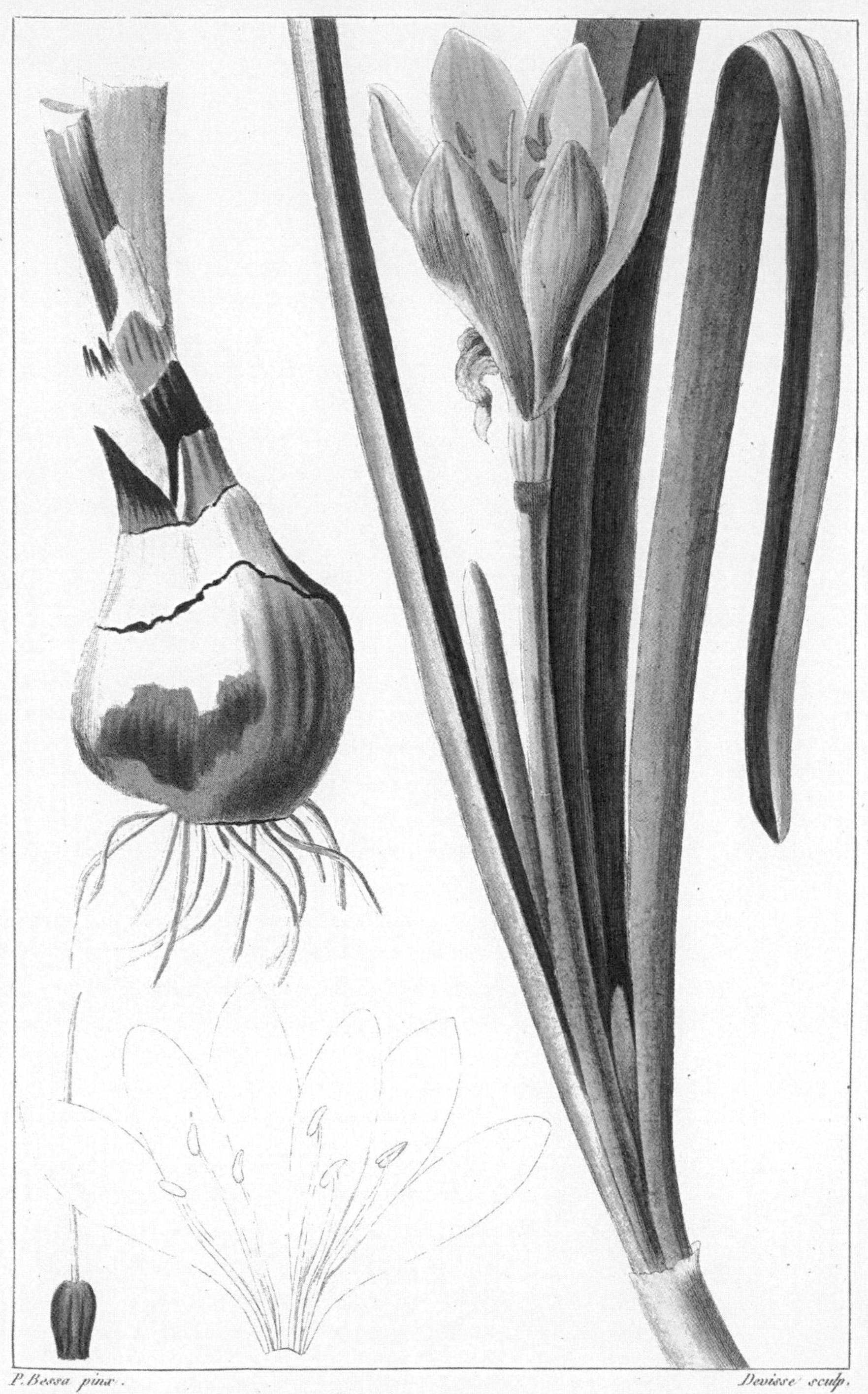

P. Bessa pinx. Devisse sculp.

Amaryllis lutea.

Opposite Winter daffodil (*Sternbergia lutea*) from *Herbier général de l'amateur* by Jean Claude Michel Mordant de Launay and Jean-Louis-Auguste Loiseleur-Deslongchamps, 1816.

raised bed suits them perfectly. In colder gardens they will also thrive at the base of a warm wall; all these are situations which provide important protection throughout the winter for the leaves, which must remain in fine fettle to build reserves for the bulb before they die down in late spring. Plants also succeed at the front of shrub or mixed borders and in short grass below trees, even standing a little dappled shade – if you can trim nearby planting to allow sun in once leaves appear in autumn, so much the better.

Sternbergia are usually available to buy as dry, bare bulbs from late summer. Before getting these into the ground, it is advisable to soak them in water overnight so that they are fully rehydrated; alternatively buy and plant potted bulbs a little later in the season just as leaves are developing. Plant bulbs 10–15cm deep and take care to avoid disturbance – winter daffodils can be slow to establish, taking a year or two to begin flowering. As with many summer-dormant plants it is easy to forget them when hidden below the soil, so ensure you label clearly. Finding planting partners is a challenge; try with other autumn-flowering bulbs such as *Colchicum* or *Nerine*, or beside small clump-forming perennials such as *Iris unguicularis*. You can allow summer annuals such as *Cerinthe major*, *Eschscholzia californica*, *Myosotis* and *Nigella* to fill space while bulbs lie dormant below as long as they are removed once the bulb stirs from its summer siesta.

OTHER AUTUMN-FLOWERING BULBS TO TRY

- ***Acis autumnalis*** (autumn snowflake) is a delightful little bulb with fine grassy foliage that dies down just before slender 10cm stalks bearing nodding white flowers arise in autumn. A delicate performer, it is ideal for a sunny rock garden. Hardy to -15°C.
- ***Crocus goulimyi*** is a true autumn-flowering crocus, often seen wild in Greek olive groves with elegant white or lilac petals held atop a slender 10cm tall tube, above fine foliage. It soon makes a nice little clump of flowers given a sunny spot. Hardy to -20°C.
- ***Colchicum autumnale*** (autumn crocus) is not a crocus at all, but a member of the lily family, producing clusters of showy white, purple or pink flowers 6cm across, large leaves appearing in spring. There are many fine species and selections of ***Colchicum*** worth growing. Hardy to -15°C.

Purpletop vervain

Verbena bonariensis

Tall and slender, with long-lasting heads of butterfly-attracting flowers, this popular herbaceous perennial is perfect for a sunny, dry position and will freely self-seed.

Opposite Purpletop vervain (*Verbena bonariensis*) illustrated by Malcolm English, 2024.

Few herbaceous perennials have seen as rapid a rise in popularity among gardeners as *Verbena bonariensis*; just a few decades ago this summer-flowering perennial was seldom encountered in cultivation, yet now it is an indispensable element in many borders in temperate and Mediterranean gardens around the world. Part of the reason for this widespread appeal is the plant's great ability to flourish in dry soil, standing considerable heat and drought, while producing profuse flat-topped heads of lightly scented pink-lilac flowers for months on end, attracting butterflies and other pollinators. Its stature and form also make it exceptionally useful. The square stems, rough to the touch, are rigid, slender and strongly upright, reaching around 1.8m tall in good conditions, while the toothed, lanceolate foliage is sparse and narrow, characteristics that often lead to the plant being described by gardeners as 'see-through', meaning it can be planted at the front of a border to add height, without blocking views of plants further back.

This plant is a native of South America, growing from Colombia to Chile, but most famously in Argentina, as reflected in the plant's Latin name (*bonariensis* meaning 'from Buenos Aires', the capital). It has become widely naturalized through self-sowing in countries such as Australia and New Zealand, while in parts of the USA it is now considered invasive, often to be found on waste ground and roadsides. Garden escapes in southern parts of the UK are now also frequent, due to its popularity and, possibly, the warming climate. It is hardy to around -10°C.

Verbena bonariensis likes a sunny place in well-drained, fertile soil best, although it even thrives in quite sandy places and frequently self-seeds into paving cracks. This free-spirited nature is part of the plant's great appeal and when happy, drifts of the plant can make a great display, swaying in the breeze and enlivened by masses of bees, butterflies and hoverflies, especially in gravel gardens or on sunny terraces where the plant is allowed a free hand. Young seedlings are easily lifted and repositioned, but older plants hate disturbance. Stems never need staking, but if positioned too close to paths the brittle growth may become damaged.

Opposite *Verbena rigida* from *Curtis's Botanical Magazine*, 1832.

Flowering generally gets going in July and continues well into autumn, individual blooms are tiny – the little purple petals held atop a pinkish tube but collected into rounded heads around 5–7 cm across. Some plants may appear fractionally pinker or more purple in flower, but generally there is little variation in colour, certainly among cultivated plants. As individual blooms fade, as summer draws to a close, removing faded heads can prolong the overall display, and at the end of the season this will also prevent excessive self-seeding.

Plants are herbaceous, but lower stems become rather woody and it is these that endure in mild, temperate climates, plants usually surviving three to five years. After flowering finishes, leave the stems for their architectural quality, or tidy by reducing down to around 30cm. The plant produces fresh shoots from the base in late spring, by which point the old stems, reduced or not, can be cut right down. In cold locations plants may not survive winter, in which case they can be treated as annuals.

Verbena bonariensis is a great mixer and will combine well with a vast array of plants. In a gravel garden, grow it beside grasses such as swaying *Celtica gigantea* or *Calamagrostis*; use it to echo the flat heads of *Achillea*; thread it through plantings of lavender, *Lotus hirsutus*, and *Pseudodictamnus acetabulosus*; or mix it with daisies of *Rudbeckia*, yellow trumpets of *Oenothera* or fiery torches of *Kniphofia*, even with Japanese anemones or between roses. The possibilities are endless.

OTHER *VERBENA* TO TRY

- ***Verbena bonariensis* 'Lollipop'**; if purpletop vervain usually grows too tall, dwarf **'Lollipop'** offers an interesting alternative, reaching around 60cm tall and as much across, forming a dome-shaped plant quite different in character. Hardy to -10°C.
- ***V. hastata*** is of similar stature to purpletop vervain, but with spires of pink, mauve or purple flowers grouped in clusters. It is also leafier and less drought-tolerant. Hardy to -10°C.
- ***V. rigida*** is a cracking, compact-growing species to around 40cm with well-branched, leafy stems bearing round heads of bright purple flowers. Spreads by runners and stands a little drought. Hardy to -5°C.

W. J. H. delt. Pub. by S. Curtis Glazenwood Essex. Jan 1. 1832

Lesser periwinkle

Vinca minor

Evergreen ground-covering plant bearing starry, usually blue flowers in spring; a vigorous and extremely tough plant that will thrive even in bone-dry, root-filled soils under trees.

Dry shade is one of the hardest situations to overcome, and the number of ornamental plants that will thrive here is limited. Among the most reliable is *Vinca minor*, a plant admired for its toughness and ability to provide evergreen ground cover, even in poor, dry, root-filled soil in both sun or shade. It spreads by producing long, runner-like stems that root into the soil as they go, forming a dense mat of shining foliage. The oval leaves are attractively glossy, especially in the spring as new growth flushes bright green. At much the same time flowers appear – these are five-petalled and propeller-shaped, indicating this is a member of the Apocynacae family. Buds emerge from leaf axils of new stems, flowers 3cm wide and usually soft blue, although there are selections in different colours. Flowering lasts several weeks through spring, forming a good early source of nectar for pollinating insects. This species is a largely European native, although it can also be found as far east as southwestern Asia carpeting woodland floors and scrambling through hedgerows as well as colonizing disturbed ground. It is now widely naturalized in many other parts of the world. Plants are hardy to around -20°C.

This periwinkle could not be easier to grow, thriving in soils from light sand to heavy clay as long as not waterlogged. It needs little maintenance once established, although after planting, improving soil with organic matter give it a good start. Growth will eventually smother most weeds, and it is a good plant to cover banks and hold back loose soil. The key is to plant in the right place: in small spaces it may outgrow its welcome.

There is considerable variety among cultivars of this plant. Flower colour varies from the rich purple of *Vinca minor* 'Atropurpurea' through to pure white *V. minor* f. *alba* 'Gertrude Jekyll'. Teaming *Vinca* with herbaceous plants is tricky as they easily outcompete neighbours – try tough shade lovers such as *Epimedium × rubrum*, *Geranium endresii*, ground-covering ivies, *Pachysandra terminalis* and drought-tolerant ferns such as *Asplenium scolopendrium*. Other lesser periwinkles to consider include 'Bowles's Variety', producing glowing lavender blue flowers; 'Azurea Flore Pleno', featuring double blue-purple blooms; and 'Illumination', which has glowing gold-splashed foliage.

Opposite Lesser periwinkle (*Vinca minor*) from *Flora Londinensis* by William Curtis, 1778.

Index

(Page numbers in *italic* refer to illustrations and captions)

Acacia 7
Acacia baileyana 10
Acacia dealbata 8–11, *9*, 10
Acacia dealbata subsp. *subalpina* 9
Acacia pravissima 10
Acanthus mollis 12–15, *13*, *15*; 'Hollard's Gold' 15; 'Rue Ledan' 15
Acanthus spinosus 15
Acis autumnalis 197
Adam's needle 136
Africa 7, 47, 87, 119, 139, 143
African daisy (*Dimorphotheca jacunda*) 86–9, *87*
Agapanthus 'Alan Street' 18; Poppin' Purple ('MP003') 18; Twister ('Ambic001') 18; 'Windsor Grey' 18
Agapanthus campanulatus 16–19, *17*
Agapanthus inapertus 18
Agapanthus praecox 18
agave 7, 21, 23, 92
Agave americana 20–23, *21*, *23*
Agave attenuata 23
Agave montana 23
Agave parryi 23
Algerian iris (*Iris unguicularis*) 138–41, *139*, *141*
Allium 'Summer Beauty' 26
Allium cristophii 24–7, *25*, *26*
Allium hollandicum 'Purple Sensation' 26, 112, 144
Allium schubertii 26, *26*
Aloe arborescens 31
Aloe ferox 31
Aloe vera 28–31, *29*, *31*
Aloiampelos striatula 31
Americas 7, 96, 187, Central 99, North 21, 41, 45, 47, 51, 181, 183, South 6, 7, 37, 55, 56, 67, 167, 187, USA 61, 103, 167, 175, 199
Ampelodesmos mauritanicus 48
Anemanthele lessoniana 48
Arctotis x *hybrida* 88
Argentina 33, 37, 61, 187
Aristaloe aristata 31
Asia 51, 61, 71, 123, 187, 195, 203
Australia 6, 7, 8, 9, 34, 47, 103, 107, 109, 167, 171, 179, 199
Austrian pine 181
autumn crocus 197
autumn snowflake 197

Barbados aloe (*Aloe vera*) 28–31, *29*, *31*
beach rose (*Rosa rugosa*) 186–9, *187*
bear's breeches (*Acanthus mollis*) 12–15, *13*, *15*
bearded *Iris* 80, 85
bell African lily (*Agapanthus campanulatus*) 16–19, *17*
Beschorneria septentrionalis 33, 34
Beschorneria yuccoides 32–5, *33*, *34*;
Beschorneria yuccoides 'Flamingo' 34
Beth's poppy 104
blue daisy (*Felicia amelloides*) 114–7, *115*, *117*
blue fescue 48
blue oat grass 48
Bougainvillea 36–9, 42
Bougainvillea 'California Gold' 39; 'Los Banos Beauty' 39; 'Pedro' 39; 'Sundown Orange' 39
Bougainvillea glabra 'Sanderiana' 36–9, *37*, *39*
Bougainvillea x *buttiana* 'Poulton's Special' 39; 'Raspberry Ice' 39
Bowden lily (*Nerine bowdenii*) 154–7, *155*, *157*
Brazil 37
bridal veil 128
bristlecone pine 181
Bupleurum fruticosum 120

Calendula officinalis 104
California fuchsia (*Epilobium canum*) 94–7, *95*, *96*
California poppy (*Eschscholzia californica*) 102–5, *103*, *104*
California tree poppy (*Romneya coulteri*) 182–5, *183*
Californian lilac (*Ceanothus arboreus*) 44–5, *45*
Cambridge Cottage, Royal Botanic Gardens, Kew 6

Campsis grandiflora 40–43, *41*, *42*; hybrid 'Madame Galen' 42
Campsis radicans 41, 42; 'Atrosanguinea' 42; 'Flava' 42
Campsis x *tagliabuana* 'Indian Summer' 42; 'Kudian' 42
Canary Island date palm 174–7, *175*, *176*
catmint (*Nepeta racemosa*) 150–53, *151*, *152*
Ceanothus arboreus 44–5, *45*
Celtica gigantea 48
Cenchrus alopecuroides 48
Cenchrus longisetus 46–9, *47*
Cenchrus orientalis 48, *48*
century plant (*Agave Americana*) 20–23, *21*, *23*
Cercis canadensis 'Forest Pansy' 53; 'Heart of Gold' 53; 'Ruby Falls' 53
Cercis chinensis 'Avondale' 53
Cercis siliquastrum 50–53, *51*, *53*; 'Bodnant' 53
Cercis siliquastrum f. *albida* 53
Cereus aethiops 56
Cereus forbesli 'Spiralis' 56
Cereus repandus 'Monstrosus' 54–7, *55*, *56*
Cestrum parqui 85
Chamaerops humilis 'Vulcano' 58–61, *59*, *61*,
Chapparal yucca (*Hesperoyucca whipplei*) 134–7, *135*, *136*
Chatto, Andrew 6
Chatto, Beth 6
Chile 103, 199
China 41, 51, 187
Chinese trumpet vine (*Campsis grandiflora*) 40–43, *41*, *42*
Cistus 7, 10
Cistus creticus 64
Cistus x *aguilarii* 64
Cistus x *argenteus* 'Silver Pink' 64
Cistus x *corbariensis* 64
Cistus x *hybridus* 64
Cistus x *incanus* 'Sunset' 64
Cistus x *pulverulentus* 64
Cistus x *purpureus* 62–5, *63*, *64*
climate change 6, 7
Colchicum autumnale 197
Colombia 199
common fig (*Ficus carica*) 122–5, *123*, *125*
common rosemary (*Salvia rosmarinus*) 190–93, *191*, *192*
Cootamundra wattle 10
Cortaderia selloana 66–9 *67*, *69*; 'Aureolineata' 69; 'Pink Feather' 69; 'Pumilla' 69; 'Sunningdale Silver' 69; Tiny Pampa ('Day1') 69
creeping broom 128
Crete 139
Croatia 143, 179
crocus 139, 197
Crocus 77, 155, 184
Crocus goulimyi 197
Cupressus sempervirens 70–73, *71*, *72*; 'Green Pencil' 72; 'Swayne's Gold' 72; 'Totem' 72
Cyclamen cilicium 77
Cyclamen coum 77
Cyclamen graecum 77
Cyclamen hederifolium 74–7, *75*, *77*

daffodils 25, 104, 112, 133, 139, 194, 197
Dalmatian iris 141
date palm *176*
Delosperma congestum 80
Delosperma cooperi 78–81, *79*, 80
Delosperma nubigenum 80
Delosperma sutherlandi 80, *80*
Dendromecon rigida 82–5, *83*, *85*
desert fan palm 61
Dimorphotheca jacunda 86–9, *87*
Dimorphotheca ecklonis 88
Duke's Garden, Royal Botanic Gardens, Kew 6
Dutch garlic 26
Dwarf fan palm (*Chamaerops humilis*) 58–61, *59*, *61*

Echeveria affinis 92
Echeveria agavoides 92, *92*
Echeveria elegans 90–93, *91*, *92*
Echeveria lilacina 92
Elaeagnus 'Quicksilver' 165
England 143
English lavender (*Lavandula angustifolia*) 142–5, *143*, *144*
Epilobium canum 94–7, *95*, *96*;'Albiflorum' 96; 'Dublin' 96; 'Olbrich Silver' 96; 'Solidarity Pink' 96
Erigeron annuus 101
Erigeron aureus 'Canary Bird' 101
Erigeron glaucus 101
Erigeron karvinskianus 98–101, *99*
Eschscholzia californica 102–5, *103*, *104*

Eucalyptus gunnii 'Rengun' (France Bleu®) 109
Eucalyptus nicholii 106–9, *107*
Eucalyptus parvula 109
Eucalyptus pauciflora 109, *109*, 110–13, *111*, *112*
Euphorbia amygdaloides susbsp. *amygdaloides* 112
Euphorbia amygdaloides susbsp. *robbiae* 110–13, *111*, 112
Euphorbia mellifera 112
Euphorbia myrsinites 112
Euphorbia rigida 112

Felicia amelloides 114–7, *115*, *117*
Felicia petiolate 117
Ferula communis 118–21, *119*, *120*
Ferula tingitana 120
Festuca glauca 48
Ficus carica 122–5, *123*, *125*
figs 125; 'Osborn's Prolific' 125; 'Panachee' 125; 'White Marseilles' 125
Flexuosa Group 157
Fountain grass (*Cenchrus longisetus*) 46–9, *47*
France 9, 75, 131, 143, 179, 195
Fremontodendron californicum 85
French lavender 144
Furcraea longaeva 34

Gazania rigens 88
Genista aetnensis 126–9, *127*, *128*
Genista lydia (Lydian Broom) 128
Genista pilosa 'Procumbens' 128
Geranium endressii 133
Geranium macrorrhizum 133, *133*
Geranium x *magnificum* 133
Geranium nodosum 130–33, *131*
ghost echeveria 92
giant fennel (*Ferula communis*) 118–21, *119*, *120*
giant oat grass 48
Greece 139, 143, 195
Greek cyclamen 77

Hawaii 47, 67, 99
Helictotrichon sempervirens 48
Hesperoyucca whipplei 134–7, *135*, *136*
honey spurge 112

Iran 25, 71
Iraq 71
Ireland 34, 87
iris 26,
Iris florentina 141
Iris foetidissima 141
Iris innominata 141
Iris pallida 141
Iris unguicularis 138–41, *139*, *141*, 197
Italian cypress (*Cupressus sempervirens*) 70–73, *71*, *72*
Italy 125, 131, 143, 163, 165, 168, 179, 195
ivy-leaved cyclamen (*Cyclamen hederifolium*) 74–7, *75*, *77*

Japanese anemone 96, 200
Japanese holly 144
Judas tree (*Cercis siliquastrum*) 50–53, *51*, *53*
Juniperus scopulorum 72
Juniperus scopulorum 'Blue Arrow' 72
Juniperus scopulorum 'Skyrocket' 72

knotted cranesbill (*Geranium nodosum*) 130–33, *131*

Latifolius Group 13, 15
Lavandula angustifolia 142–5, *143*, *144*; 'Ashdown Forest' 144; 'Loddon Pink' 144; 'Purity' 144
Lavandula x *intermedia* 'Grosso' 144; 'Olympia' ('Downoly') 144; 'Walberton's Silver Edge' ('Walvera') 144
Lavandula latifolia 144
Lavandula stoechas 144
lavender 10, 45, 53, 61, 64, 115, 139, 141, 143, *144*, 151, 161, 173, 192, 200, 203
Leonotis 7
Leonotis leonurus 146–9, *147*, *149*
Lespedeza thunbergii 85
lesser periwinkle (*Vinca minor*) 202–3, *203*
lily *18*, 197
Limnanthes douglasii 104
lion's tail (*Leonotis leonurus*) 146–9, *147*, *149*
Lophocereus marginatus 56
love-in-a-mist 104
Lydian broom 128

Mediterranean Garden, Royal Botanic Gardens, Kew 6
Mexican fleabane (*Erigeron karvinskianus*) 98–101, *99*
Mexican lily (*Beschorneria yuccoides*) 32–5, *33*, *34*
Mexican snowball (*Echeveria elegans*)

90–93, *91*, *92*
Mexico 21, 23, 33, 45, 51, 56, 61, 83, 91, 95, 99, 103, 135, 167, 183
Middle East 122, 123, 143, 151, 187
miniature date palm 176
Morocco 179
moss-rose purslane 104
Mount Etna broom (*Genista aetnensis*) 126–9, *127*, *128*
mountain pine 181
Mrs Robb's bonnet (*Euphorbia amygdaloides* susbsp. *robbiae*) 110–13, *111*, 112
Myrrhis odorata 120

narrow-leaved black peppermint (*Eucalyptus nicholii*) 106–9, *107*
Nepeta 'Six Hills Giant' 152
Nepeta grandiflora 'Dusk till Dawn' 152
Nepeta nepitella 152
Nepeta racemosa 150–53, *151*, *152*
Nepeta sibirica 'Souvenir d'André Chaudron' 152
Nerine bowdenii 154–7, *155*, *157*
Nerine undulata (Flexuosa Group) 'Alba' 157
Nerine undulata 'Seaton' 157
Nerium 'Alsace' 160; 'Angiolo Pucci' 160; 'Madame Allen' 160; 'Maresciallo Graziani' 160; 'Oasis' 160; 'Rubis' 160; 'Roseum Plenum' 160
Nerium oleander 158–61, *159*, *160*
New Zealand 9, 13, 33, 67, 99, 199
Nigella damascena 104
Norfolk 143

old man prickly pear 168
Olea europaea 162–5, *163*, *165*
oleander (*Nerium oleander*)158–61, *159*, *160*
olive (*Olea europaea*) 162–5, *163*, *165*
Oman 29
Opuntia englemanii var. *cuija* 168
Opuntia erinacea 168
Opuntia ficus-indica 166–9, *167*, *168*
Opuntia humifusa 168
organ pipe cactus 56
Osteospermum jucuandum 86–9, *87*, *88*
oven's wattle 10

Pacific Coast iris 141
pampas grass (*Cortaderia selloana*) 66–70, *67*, *69*
Papaver lecoqli 'Albiflorum' 104
Pelargonium 'Attar of Roses' 173; 'Prince of Orange' 173
Pelargonium capitatum 170–73, *171*, *173*
Pelargonium tomentosum 173
Pennisetum alopecuroides 48
Pennisetum orientalis 48, *48*
peppermint geranium 173
Peru 37, 55, 56
Peruvian apple cactus (*Cereus repandus*) 54–7, 55, 56
pheasant's tail grass 48
Phillyrea latifolia 165
Phlomis russeliana 149
Phlomoides tuberosa 'Amazone' 149
Phoenix canariensis 174–7, *175*
Phoenix dactylifera 176
Phoenix roebelenii 176
Pinus aristata 181
Pinus mugo 181
Pinus nigra 181
Pinus pinea 178–81, *179*, *181*
poached egg plant 104
Portaluca grandiflora 104
Portugal 179
pot marigold 104
potato vine 85
prickly pear (*Opuntia ficus-indica*) 166–9, *167*, *168*
purple tulips 15
purpletop vervain (*Verbena bonariensis*) 198–201, *199*, 200

red sunflower 104
red-leaved rose 189
Retama monsperma 128
roast-beef plant 141
Romneya 7, 184
Romneya 'White Cloud' 184
Romneya coulteri 182–5, *184*
Romneya coulteri subsp. *trichocalyx* *183*, 184
Rosa 'Fru Dagmar Hastrup' 189; 'Hansa' 189
Rosa rugosa 186–9, *187*, *189*; 'Alba' 189
rose-scented geranium (*Pelargonium capitatum*) 170–73, *171*, *173*
rosemary 191, 192
Royal Botanic Gardens, Kew 6, 7

Salvia rosmarinus 190–93, *191*, *192*; 'Foxtail 192; 'Majorca Pink'

192; 'Miss Jessup's Upright' 192; 'Roman Beauty' 192
Schubert's allium 27, *27*
Scotland 155
Siberia 187
silver wattle (*Acacia dealbata*) 8–11, *9*, *10*
small leaved gum 109
Solanum crispum 85
South Africa 13, 17, 55, 78, 79, 114, 115, 147, 154, 155, 171, 179
Spain 143, 163, 179, 195
Spanish dagger 136
star of Persia (*Allium cristophii*) 24–7, *25*, *26*
Steonocereus griseus 56, 194–7, *195*, *197*
Sternbergia lutea 194–7, *195*, *197*
stone pine (*Pinus pinea*) 178–81, *179*, *181*
sun rose (*Cistus* x *purpureus*) 62–5, *63*, *64*
sweet cicely 120
Syria 179

Tamarix tetrandra 165
Tithonia rotundifolia 104
trailing ice plant (*Delosperma cooperi*) 78–81, *79*, *80*
tree poppy (*Dendromecon rigida*) 82–5, 83, 85
Trithrinix campestris 61
Turkey 75, 111, 139, 151, 179
tulips 15, 25, 104, 112, 119, 120
Turkmenistan 25

United Kingdom 6, 9, 17, 67, 83, 111, 112, 171, 176, 183, 187, 199

Verbena bonariensis 'Lollipop' 198–201, *199*, *200*
Verbena hastata 200
Verbena rigida 200, *200*
Vinca minor 202–3, *203*

Washingtonia filifera 61
willow-leaved jessamin 85
winter daffodil (*Sternbergia lutea*) 194–7, *195*, *197*

Yucca 135, 136
Yucca filamentosa 136
Yucca gloriosa 136
Yucca rostrata 136

Zauschneria californica 94–7, *95*, *96*

Credits

The publishers would like to thank the following sources for their kind permission to reproduce the pictures in this book.

All other images are taken from the Library and Archives collection of the Royal Botanic Gardens, Kew.

BIODIVERSITY HERITAGE LIBRARY: Harvard University Botany Libraries: 153

© CHRISTABEL KING: 98

MISSOURI BOTANICAL GARDEN, ST. LOUIS: © Peter H. Raven Library: 97